Hydrocolloids in food product development

Albert Monferrer
Claudia Cortés
Núria Cubero
Laura Gómez

Hydrocolloids in food product development

Albert Monferrer
Claudia Cortés
Núria Cubero
Laura Gómez

Originally published in Spanish by BDN Ingeniería de Alimentación. S.L. through viveLibro with the title *Modiyfing the Texture of Food: Handbook of Hydrocolloids*. The authors prepared the translation, which was initially reviewed and corrected by Maya Rieder

CRC Press
Taylor & Francis Group
6000 Broken Sound Parkway NW, Suite 300
Boca Raton, FL 33487-2742

© 2020 by Taylor & Francis Group, LLC
CRC Press is an imprint of Taylor & Francis Group, an Informa business

No claim to original U.S. Government works

Printed on acid-free paper

International Standard Book Number-13: 978-0-367-89552-5 (Hardback)

This book contains information obtained from authentic and highly regarded sources. Reasonable efforts have been made to publish reliable data and information, but the author and publisher cannot assume responsibility for the validity of all materials or the consequences of their use. The authors and publishers have attempted to trace the copyright holders of all material reproduced in this publication and apologize to copyright holders if permission to publish in this form has not been obtained. If any copyright material has not been acknowledged please write and let us know so we may rectify in any future reprint.

Except as permitted under U.S. Copyright Law, no part of this book may be reprinted, reproduced, transmitted, or utilized in any form by any electronic, mechanical, or other means, now known or hereafter invented, including photocopying, microfilming, and recording, or in any information storage or retrieval system, without written permission from the publishers.

For permission to photocopy or use material electronically from this work, please access www.copyright.com (http://www.copyright.com/) or contact the Copyright Clearance Center, Inc. (CCC), 222 Rosewood Drive, Danvers, MA 01923, 978-750-8400. CCC is a not-for-profit organization that provides licenses and registration for a variety of users. For organizations that have been granted a photocopy license by the CCC, a separate system of payment has been arranged.

Trademark Notice: Product or corporate names may be trademarks or registered trademarks, and are used only for identification and explanation without intent to infringe.

Library of Congress Cataloging-in-Publication Data
Names: Jiménez Cortés, Claudia, author.	Cubero, Nuria, author.	Gómez, Laura, 1982- author.	Monferrer, Albert, author.	BDN Food Ingeniería de Alimentación, issuing body.
Title: Hydrocolloids in food product development / BDN Food Ingeniería de Alimentación: Claudia Cortés, Núria Cubero, Laura Gómez, Albert Monferrer.
Description: Boca Raton, FL : CRC Press, [2020]	Includes bibliographical references.
Identifiers: LCCN 2019048325 (print)	LCCN 2019048326 (ebook)	ISBN 9780367895525 (hardback)	ISBN 9781003019862 (ebook)
Subjects: LCSH: Hydrocolloids.	Food additives.
Classification: LCC TP456.H93 J55 2020 (print)	LCC TP456.H93 (ebook)	DDC 664/.06--dc23
LC record available at https://lccn.loc.gov/2019048325
LC ebook record available at https://lccn.loc.gov/2019048326

Visit the Taylor & Francis Web site at
http://www.taylorandfrancis.com

and the CRC Press Web site at
http://www.crcpress.com

Originally published in Spanish by BDN Ingeniería de Alimentación S.L. through viveLibro with the title *Modifying the Texture of Food: Handbook of Hydrocolloids*. The authors prepared the translation, which was initially reviewed and corrected by Maya Rieder.

AUTHORS

Claudia Cortés

Obtained an Agricultural Technical Engineering degree, specialized in agricultural and food industries from ESAB (Escola Superior d'Agricultura de Barcelona) (UPC, 2009), and graduated in Food Science and Technology from the department of Veterinary studies in the Universitat Autònoma de Barcelona (UAB, 2011).

Started working in BDN Ingeniería de Alimentación (Barcelona) in 2012, cooperating technologically with departments of R&D for companies coming from various sectors in the food industry.

Núria Cubero

Obtained an Agricultural Technical Engineering degree, specialized in agricultural and food industries from ESAB (Escola Superior d'Agricultura de Barcelona) (UPC, 2000), and graduated in Food Science and Technology by the school of Veterinary studies at the Universitat Autònoma de Barcelona (UAB, 2004).

Started working in BDN Ingeniería de Alimentación (Barcelona) in 1999, cooperating technologically with departments of R&D for companies coming from various sectors in the food industry.

Holder of three patents developed at BDN. Since 2007 she collaborates as a professor in the food additive block in the Master for Technology, Control and Safety in the Food Industry (MTCA) in the Centro de Estudios Superiores de la Industria Farmacéutica (CESIF). She is associate professor in the degree program for Culinary Science and Gastronomy, at Universitat de Barcelona (UB).

Laura Gómez

Graduated in Biology from the Universitat de Barcelona (UB, 2005), graduated in Food Science and Technology from the Universitat de Barcelona (UB, 2008) and obtained a Master's Degree in Development and Innovation in Food (UB, 2009).

Started working in BDN Ingeniería de Alimentación (Barcelona) in 2007, cooperating technologically with departments of R&D for companies coming from various sectors in the food industry.

Albert Monferrer

Graduated in Veterinary Medicine (Universidad de Zaragoza, 1984) and obtained a Master's Degree in Food Biotechnology (UPC, 1989).

Started working in BDN Ingeniería de Alimentación (Barcelona) in 1992, cooperating technologically with departments of R&D for companies coming from various sectors in the food industry.

Holder of six patents developed at BDN.

Since 2005 he has been associate professor in the Department for Nutrition, Food Science and Gastronomy in the school of Pharmacy and Food Science at the Universitat de Barcelona for the degree of Food Science, degree of Culinary Science and Gastronomy and Master for Development and Innovation in Food.

For years he has been collaborating as a professor in the Master of Nutrition and Metabolism in Universitat Rovira i Virgili, Master of Food Safety in the Official College of Veterinarians of Madrid and in the Higher Diploma of Food, Nutrition and Public Health in Instituto Carlos III of Madrid.

PREFACE

According to Wikipedia (Spanish version, May 2016), "a preface is, in literature, an introduction and presentation text, located at the beginning of a book". This manual isn't literature itself, but the preface will also serve to introduce the book and, in addition, we will put it in the right place. At the beginning.

It is also interesting, following the writings of Wikipedia, the beginning of the "Antecedents" chapter, which says "the ancients put prefaces at the start of his works". It certainly comes to be that. BDN just made 29 or almost 30 years of existence, and for the medium life of many companies, we already are ancient.

But... what and who is BDN?

We will use the preface to introduce ourselves, and what better way to provide answers to the eternal questions about us, where do we come from? where are we going?

Who are we?

We are a group of professionals that offer technical consultancy to companies related to the manufacturing of food, especially in all that refers to the formulation, use of ingredients and additives, processes and development of new products.

Working throughout these 29 or almost 30 years for various sectors and making very different formulations has caused, little by little, our understanding and we have applied the technological properties of many ingredients and additives. Through these years we have been asked to participate in numerous courses, lectures, seminars... to explain the how and the why of these products in a very practical way. Talks about hydrocolloids have been, possibly, the more successful ones. For this reason, we have decided to write this manual.

This manual is not intended to be a Bible or an encyclopedia. Nothing further from reality. Our intention was to put within reach of the technicians of the companies, the students and the chefs that are using additives, a concise and schematic recompilation of the main technological characteristics of hydrocolloids.

In this manual we have collected basic technical information, which you can find much more expanded and maybe, better explained, in many textbooks. But we have also included experiences, personal comments and practical information, which we believe may be useful to all those who go into the world of food texture.

Where do we come from?

BDN wouldn't exist if its founder, Jordi Villalta, wasn't be as he is. His friends have defined him as "crazy wise", "pou de ciència (well of science)", "mestre (master)" and many more. The truth is that Jordi is special. Faithful to his principles, frank and noble, practical and with different ideas. When you ask his opinion about how to solve a problem in a formulation, he usually gives you three options: one that is typical and is found in all the books; another that is impossible, improbable or a nonsense, and the third, which tends to be a stroke of genius or, at least, a different solution to the usual.

Jordi has created school. Not only at BDN, also among the technicians of companies that have known and worked with him.

Thanks, Jordi, for being there.

Where are we going?

We will go where the customers, the market and the food innovation take us. Our intention is to continue working with companies, solving the everyday problems, counseling or providing different points of view and helping companies innovate.

We want to continue working next to the technical departments of the companies, grow and expand.

We also want to continue spreading and sharing the knowledge of the use of ingredients and additives either through talks or, perhaps, with some other manual like this. Thank you for joining us on our voyage.

The authors

Barcelona, July 2019

INDEX OF CONTENTS

Chapters guide 1

HYDROCOLLOIDS AND TEXTURE 7

INTRODUCTION TO HYDROCOLLOIDS 25

E-400 ALGINATES 37

E-406 AGAR 49

E-407 CARRAGEENAN 57

E-410, E-412, E-417, E-427 GALACTOMANNAN 67

E-413 TRAGACANTH GUM 81

E-414 GUM ARABIC 87

E-415 XANTHAN GUM 93

E-416 KARAYA GUM 101

E-418 GELLAN GUM 105

E-425 KONJAC 113

E-440 PECTIN 123

E-460i MICROCRYSTALLINE CELLULOSE 133

E-461, E-464, E-466 MODIFIED CELLULOSES 139

E-1204 PULLULAN 157

IDENTIFICATION AND GRAS STATUS 161

BIBLIOGRAPHY 163

0 Chapters guide

WHAT WILL BE FOUND IN EACH CHAPTER?

TITLE (E-XXX FOOD ADDITIVE NAME)

It is the name of the food additive and its number E. All the permitted additives are common to all European Union and its use in food is collected in Regulation 1129/2011 from 11 November 2011 amending Annex II to Regulation (EC) no 1333/2008 of the European Parliament and of the Council to set a list of food additives of the Union.

It also includes its CAS number (Chemical Abstracts Services). This number is assigned by the American Chemical Society and identifies each chemical compound described.

Other denominations of the additive described in literature are also collected.

Example: agar

E-406 Agar (Agar, CAS Number: 9002-18-0)

Other denominations: agar agar agar, kanten.

Origin

Describes where it comes from or how the additive is obtained. It is complemented with a diagram of the obtainining process.

Composition and chemical structure

It explains the composition and basic chemical structure of the additive, i.e., the type of chemical compound, the proportion and types of radicals that they can form and, therefore, variations that may have additive.

Also an image of the molecule or molecules that form it is attached.

Properties of solutions

Solubility and solutions preparation
From a 1% solution of the additive in water their ability and solubility characteristics are determined. It is complemented by the following table:

Example: alginate

Table 1. Solubility and solutions preparation / +++ (high) ++ (middle) + (low)

Solubility in cold water	Solubilization temperature	Premix	Shear force	Air incorporation	Increase viscosity of medium	Precipitates with alcohol	Ion sequestrant
+++	cold	+++	+++	++	+	+	+++

- **Solubility in cold water:** the additive dissolves well in cold water? It makes lumps? It is punctuated with +++, if it dissolves well and does not make lumps and +, if the dissolution is bad and makes lumps.

- **Temperature of solubilization:** the additive has to be dissolved in hot or cold water?

- **Premix:** does it need to be previously mixed with other ingredients to facilitate its subsequent dispersion in water and avoid the formation of lumps? +++, if it needs to premix with other ingredients and +, if the premix does not influence.

- **Shear force:** does it take intense mechanical work to properly dissolve the additive in water? +++, if needed, and +, if not needed.

- **Air incorporation:** does mechanical work cause the formation of foam in the solution? +++, if a lot of air will be incorporated in the solution, and +, if there will not be air incorporation.

- **Increase in viscosity of the medium:** Does the solubilisation of the additive modify the viscosity of the medium in cold conditions? +++, causes a great increase of the viscosity; +, has no influence.

- **Precipitates with alcohol:** is it possible to dissolve the additive in a alcoholic medium? It is made from a 1% solution of the additive in a medium with 40% of alcohol and its behavior is observed.

- **Ions sequestrant:** would the additive react to ions present in the medium? This column appears only when the use of sequestrant is needed for the correct solution of the additive.

Factors affecting the properties of solutions

There are listed the factors that can influence the behavior of the additive on the viscosity. These features are contained in the following table:

Example: galactomannan

Table 4. Factors affecting the properties of solutions

Factors	Effects on viscosity	Observations
Molecular weight/ polymerization degree	Directly proportional	
Concentration	Directly proportional	
Temperature	Inversely proportional	An increase in temperature causes a decrease in viscosity
pH	Acid pH hydrolyze the molecule	Viscosity decreases in an acidic environment
Ionic charge	Does not affect	
Mechanical work	Pseudoplastic behavior	It decreases the viscosity as it increases the shear force

1. **Molecular weight / polymerization degree:** does it affect the molecular weight of the additive on the viscosity of the medium? In the majority of the additives it is directly proportional, i.e., a higher molecular weight, higher viscosity of the medium.

2. **Concentration:** does concentration affect the viscosity of the medium? The majority is directly

proportional, i.e., to greater concentration of additive, higher viscosity of the medium.

3. **Temperature:** does increasing the temperature of the medium affect viscosity? The effect can be directly proportional (more temperature, more viscosity) or inversely proportional (more temperature, less viscosity).

4. **pH:** is the additive stable to pH changes of the medium? Do these pH changes affect the viscosity of the medium?

5. **Ionic charge:** do the present ions in the medium interact with the additive and affect the viscosity of the medium? The determination of this factor is done under normal conditions of use of the additive, not in specific cases of highly concentrated products (example: brine).

6. **Mechanical work:** has the additive pseudoplastic, thixotropic or Newtonian behaviour?

7. **Other:** notable specific factors are included for an additive in concrete and having effect on viscosity.

Functionality

As a general rule, this section explains the gelling ability of an additive. But in cases when the additive does not have functionality as a gelling agent, e.g., the case of galactomannans or the xanthan gum, this section describes the function as a thickener.

Gelation mechanism

A 1% solution of the additive is forced to make a gel and it explains the mechanism to make the molecules of the additive react to produce the gelation.

It is complemented by a scheme of solubilization and functionality of the additive.

Form of preparation

Describes the process that produces the gelation, step by step, to form a gel from a 1% solution of the additive.

Example: konjac

Thermostable gel: solubilize and hydrate well the konjac, get to a temperature above 85°C and hold for 30 minutes with continuous agitation. Stop agitation and leave to cool down to room temperature. Then add the weak base (for example, K_2CO_3) until reaching a pH of 9 or higher to deacetylate the molecule, and mix well. Begin to heat the mixture again to 85°C for 2h without agitation. Leave to cool.

Characteristics of gels

This describes the type of gels that can form an additive and their characteristics and also the factors that determine the formation of these gels. All this is collected in the following table:

Chapters guide

Example: gellan

Table 3. Characteristics of gels

Type of gellan	Low acetylation	High acetylation
Gel type	Firm and crisp	Soft and elastic
Syneresis	Yes, only in the cut	No
Temperature stability	Thermostable in the presence of Ca^{2+}. Thermorreversible with K^+ and in milk, hysteresis	Thermorreversible, no hysteresis

Factors affecting the properties of gels

It explains the factors to consider during the formation of the gel (temperature, pH, ions, ...).

Synergies or incompatibilities

It describes the different synergies of the additive with other additives or ingredients to get different functions.

Examples of applications in food industry

There is an enumeration of different functions of the additive and examples of lists of ingredients where the additive is used, which is highlighted in bold.

Example: metyl cellulose

- Gelling agent in hot conditions for baked or fried products

> **VEGGIE BURGER**
>
> Ingredients: soya beans, water, soy protein, onion, tomato concentrate, sunflower oil, salt, sugar, wheat fiber, onion juice concentrate, gelling agent (**methyl cellulose**)

Throughout the chapters three types of boxes will offer different facts related to the additive:

They give extra information about the additive in question which is not collected in any section of the chapter

Extend or complement the chapter concerning information to facilitate understanding

They explain curiosities of the additive in question

1 HYDROCOLLOIDS AND TEXTURE

Definition of texture

When evaluating food stuff organoleptically, four main characteristics stand out:

- **Appearance.** Evaluated by sight. Includes color, size, shape, shine...

- **Aroma.** Perceived by several chemical receptors located in the oral cavity, nose and retro nasal olfactory area. Comprised of two complementary concepts:

    o **Taste.** Picked up by the taste buds in the tongue and processed by the brain. Includes the five basic tastes (sweet, salty, acid, bitter and umami) and their variations and combinations.

    o **Scent.** Picked up by the chemoreceptors of the nasal epithelium and processed by the olfactory system of the brain. These receivers collect stimuli generated from gases, vapors or particles in suspension released by foods.

- **Sound.** Evaluated by the ear. It includes auditory sensations which inform about features like crispy, pasty or adherents that through sounds produced by chewing food. These sounds complement the following feature, texture.

- **Texture.** Consists of physical stimuli collected by touch receptors located on fingers, lips, tongue, and generally in oral cavities. They can be static (a simple feeling to the touch of your fingers, lips or tongue) or dynamic (collected during cutting, chewing and swallowing food).

Texture, in some cases, can interact with the smell to affect the consistency and stickiness of food and, by that, allowing it to stay more or less time in a person´s mouth.

The main function of hydrocolloids is to modify food texture when hydrated, to interact with the water, which will solubilize them, and change the properties of the water flow. But food texture does not solely depend on the use of hydrocolloids. In fact, all of the ingredients included in the formula or recipe will influence its texture (figure 1). For instance:

- Water/solids ratio
- Percentage of fat or oil, and type
- Addition of starches, flour...
- Use of proteins that can be denatured by heat
- Addition of vegetable fibers
- Formation of emulsions and foams
- Other ingredients in the formula and their percentage
- Preparation process (roasted, fried, boiled...)

Even the visual appearance of the product suggests if it will be hard, crisp or soft depending on its more or less brown in color, or smooth or rough surface, or wet or dry appearance...

Verbally describing the texture of food is complicated because there is an important component of subjectivity. Every person characterizes food differently. For instance, what seems hard to a person may seem appropriate or even soft to another. For this reason, it is necessary to establish a common vocabulary and a standard way of evaluating the texture of food. Kramer (1964) (Table 1) and Lamond (1977) (Table 2) established the foundations of the sensory analysis of texture that has been later normalized by the standard UNE - EN ISO 8586:2014

Hydrocolloids and texture

In relation to food texture, Müller (1969) proposed separating this term in two different concepts (Table 3):

- Rheology: branch of physics that describes the physical and mechanical properties of materials.

- Haptaestesis: (from the Greek "feeling" or "touch") branch of psychology that deals with the perception of the mechanical properties of materials. It applies mainly to food and cosmetic products.

The relation between rheological parameters and the sensations that they produce (haptaestesis) is studied in psychorheology.

Taking into consideration the difficulty to express and define words for food texture, and that there is a direct relation with the physics, the ideal would be to measure and quantify the attributes of texture of fluids (viscosity) and solids (characteristics of the gel). Rheology is the part of physics that studies these phenomena and how to measure them.

Table 1. Texture attributes (Adapted from Kramer (1964))

Textural characteristics		
Mechanical	**Geometrical**	**Composition**
PRIMARY		
Hardness	Fibrousity	
Cohesiveness	Graininess	Moisture
Springiness	Sponginess	Dryness
Adhesiveness	Crystallinity	Oiliness
Viscosity	Flexibility	Greasiness
SECONDARY	Friability	Sandiness
Fracturability	Turgidity	Muddy
Chewiness	Roughness	
Gumminess		
Stickiness		

Table 2. Texture attributes (Adapted from Lamond (1976))

Characteristics	Definitions
Hardness	Physical: Required force to achieve a deformation or penetration in a product.
Hardness	Sensorial: Required force to compress solids between the teeth and semi-solids between the tongue and hard palate.
Cohesiveness	Physical: Degree of deformation of a material before it breaks.
Cohesiveness	Sensorial: Amount of compression between the teeth that can be applied before the substance breaks.
Viscosity	Physical: Unit that describes a fluid's resistance to flow.
Viscosity	Sensorial: Required force to pass a liquid from a spoon over the tongue.
Elasticity or Springiness	Physical: Recuperation rate of a material to return its original condition after a force is applied.
Elasticity or Springiness	Sensorial: Degree to which a product returns to its original shape after being compressed between the teeth.
Adhesiveness	Physical: Necessary force to beat the forces of attraction between the surface of the food and that of the materials of which it is in contact.
Adhesiveness	Sensorial: Required force to remove the material stuck to the mouth or hard palate when consumed.

Table 3. Texture attributes (adapted from Müller (1969)

Comparison of physical measures vs. human perception		
Light	Optics (physical)	Reflection Refraction Light waves
Light	Vision (perception)	Size Color Form
Texture	Rheology (physical)	Young's modulus Module of shear Poisson coefficient Viscosity Elasticity
Texture	Haptaestesis (perception)	Mouthfeel Hardness Chewiness Gumminess Adhesiveness

Hydrocolloids and texture

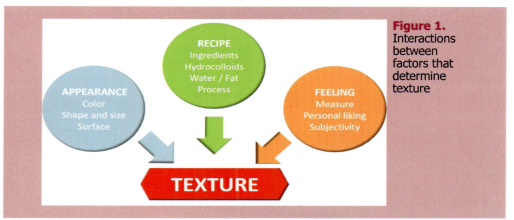

Figure 1. Interactions between factors that determine texture

Rheology

Rheology is the part of physics that studies the flow and deformation of materials subjected to the action of an external force that usually corresponds to a mechanical work.

A body subjected to a tangential force (known as *shear force* or *shear* and that is represented by the Greek letter σ) will resist deformation depending on intermolecular cohesion forces between its components, causing different behaviours:

- If cohesion forces are strong enough and greater than the tangential force, the body is not deformed (solid).

- If the cohesion forces are hard but similar or slightly smaller than the tangential forces, the body is deformed or broken (solid).

- If the cohesion forces are weak and lower than the tangential forces, the body deforms and flows (fluid).

When applying a shear force on a body, the layers on which the force is applied move faster than the surrounding (figure 2). Occasionally, this phenomenon occurs when a certain shear speed is applied. This is represented by the Greek letter ẏ (Table 4).

In the case of solids, a deformation occurs. Taking the force off can result in a return to the initial position and form (elasticity) or permanently losing the shape or breaking (deformation).

In the case of fluids, a movement or flow occurs in the different layers at different speeds, since each layer drags the bottom layer by friction. When force is withdrawn, the fluid can return to its original shape (if it stays in the same container) or it can take the form of a new recipient that contains it.

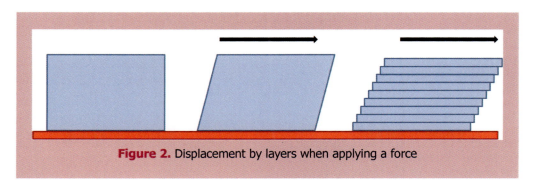

Figure 2. Displacement by layers when applying a force

Handbook of hydrocolloids

Table 4. Magnitudes, symbols and units in the International System

Magnitude	Symbol	Unit
Shear force	σ	Pa
Shear speed	$\dot{\gamma}$	s^{-1}
Viscosity	η	Pa·s

However, there is not always a very clear line between what is solid or what is fluid. Some very "thick" fluids (viscous) seem to behave like soft solids (elastic) and vice versa. For this reason, these are called viscoelastic materials, since, in many cases, they share properties of fluids and solids. Therefore, in order to understand better the behaviours, the rheology of both materials is studied separately.

There are other factors that will influence rheology of materials apart from the mechanical work to which they are subjected. The most important are the temperature and concentration. In order to understand the changes due to variations of mechanical work, you should assume that concentration and temperature remain constant.

Rheology in fluids

Dynamic viscosity

Viscosity is the opposition that shows a fluid to deform and move when a tangential force is applied. Therefore, the concept of viscosity can only be applied to fluids in motion and represents the relationship between tangential strength (shear stress or shear force) and the caused gradient of speed. This concept of viscosity (represented by the Greek letter μ) is known as absolute viscosity or dynamic viscosity. This kind of viscosity is measured by rotational viscometers or falling ball viscometers.

In the International System (SI) the unit of measure for the dynamic viscosity is the Pascal per second (Pa·s), which corresponds to 1 N·s/m^2 or 1 kg/(m.s).

In the Cegesimal System (CGS) the unit of measure for the dynamic viscosity is the poise (P) which corresponds to 1 g/(s·cm) or 0.1 Pa·s. In food science its submultiple centipoise (cP) is commonly used and is taken as reference since water at 20°C has a viscosity of 10,020 cP.

Kinematic viscosity

Another measure of viscosity is kinematic viscosity, which is represented by the v and corresponds to the ratio of dynamic viscosity and density of the fluid. This type of viscosity is measured with Ford viscosity cups and alike.

Kinematic viscosity is the ratio of dynamic viscosity and density. The SI unit is m^2/s. In the CGS unit is the stoke (St) (Table 5).

Apparent viscosity

The term "apparent viscosity" is often used in liquids whose dynamic viscosity changes depending on the applied shear force or time during which it applies, meaning that viscosity can vary according to the conditions of measurement.

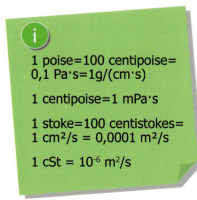

1 poise=100 centipoise= 0,1 Pa·s=1g/(cm·s)

1 centipoise=1 mPa·s

1 stoke=100 centistokes= 1 cm²/s = 0,0001 m²/s

1 cSt = 10^{-6} m²/s

Hydrocolloids and texture

Table 5. Characteristics of absolute and kinematic viscosity

Characteristics	Absolute viscosity	Kinematic viscosity
Unit	Pascal· Second (Pa·s) In food science the centiPoise (cP) is commonly used: cP = mPa·s	centiStoke (cSt)
Symbol	µ Relation between shear force and speed gradient	ν Relation between dynamic viscosity and density ν = µ/ρ
Measure equipment	Rotational viscometer Falling ball viscometer	Ford viscosity cup

Behavior of fluids according to their viscosity

Not all fluids, appearing to be more or less viscous, behave in the same way when a shear force is applied to them. In the majority of circumstances, its viscosity will remain stable, while in others, the viscosity will seem to change depending on the shear force applied or the time that this force is being applied (figure 3). These variations will have great importance in the behavior of the fluid during the manufacturing process, when being transferred from one tank to another, or at the time of packaging. Knowing which the behavior of viscous liquids will be, variables such as temperature, shear force, or application time can be changed to achieve, momentarily, a more or less viscous product. This fact will facilitate, for instance, operations as transfer, filling or transportation.

- **Newtonian fluids:** Newtonian fluids are fluids whose viscosity is constant at a given temperature and does not depend on the shear force or the time during which it is applied. These fluids meet Newton's law in such a way that the relationship between the strength of shear (σ) and shear speed (γ) is proportional and the viscosity is constant in all the velocity gradients of shear values. Therefore, the viscosity is a characteristic parameter of this fluid.
The best example of a Newtonian fluid is water. It is commonly thought that Newtonian behavior appears in low viscosity liquids, such as milk, coffee, juices, soft drinks, oils... but also high viscous liquids as glycerin, sugar syrups and honey are Newtonian fluids.

- **Non-Newtonian fluids:** most fluids in food are non-Newtonian fluids, and their viscosity varies depending on the speed of shear or, in some cases, the time of maintaining this speed of shear.

    o Depending on shear strength:

      - Pseudoplastic fluids: fluids whose viscosity decreases by increasing the shear strength. When shear is stopped, the fluid recovers its initial viscosity.
      It is interesting to take this effect into consideration when it comes to transferring or circulating a very viscous fluid. If mixing blades or recirculation pumps are used at a higher shear rate, it will be easier to move the fluid in the circuit.
      Hydrocolloids usually provide this type of feature to fluids.

      - Dilatant fluids: fluids whose viscosity increases by increasing shear strength.

When shear is stopped, the fluid recovers its initial viscosity.

These are not usual fluids in food. They appear in fluids that have a large amount of insoluble suspended particles. An example of this behavior are very concentrated suspensions of uncooked starch (Internet videos that illustrate this phenomenon can be found). Another example, in this case non-food, is the sand on the seashore. When walking on this sand, our footprint is marked since we sink into the sand and water mixture, but if we run and exert a greater shear strength, footprints are not marked as sand increases its viscosity.

- o Depending on the time of shear application

    - Thixotropic fluids: fluids whose viscosity decreases whose viscosity decreases with time by applying a constant shear effort. When shear is stopped, the fluid recovers its initial viscosity.
    Some hydrocolloids cause this type of flow. Food products such as stirred yoghurt or baby food purées have thixotropic behavior.
    In some rare cases, the initial viscosity does not recover. This phenomenon is named irreversible thixotropy or rheomalaxis. This phenomenon appears in some concentrated juices.

    - Rheopectic fluids: The viscosity of the fluid increases with the time of application of a constant shear effort. Also referred to as anti-thixotropy.
    It is not a common phenomenon in food products.

- **Yield stress fluids:** also known as Bingham plastics. These are fluids that do not flow spontaneously and need to be affected by a force or pressure exceeding a threshold effort. From that moment it begins to flow and, generally, behaves like a Newtonian fluid.
Some examples of foods that require this initial effort threshold are ketchup, mustard, mayonnaise...

The variation on viscosity, mainly its decrease, is due to a reorientation, or deformation of macromolecules, or to the breaking of agglomerates of particles (figure 4). This new conformation causes a minor interaction between molecules and facilitates their movement in the fluid.

- Rigid linear macromolecules are oriented in the direction of the flow.

- Macromolecules form a ball of connected molecules that are partially unpacked and orient themselves in the direction of the flow.

- Spheroidal macromolecules deform and acquire more hydrodynamic forms.

- Aggregates or clusters of particles are scattered in individual components.

Usually, when the liquid stops flowing, macromolecules return to their initial formation and viscosity is recovered.

Table 6 summarizes the most often used systems in food industry to measure or compare the viscosity of fluids.

Hydrocolloids and texture

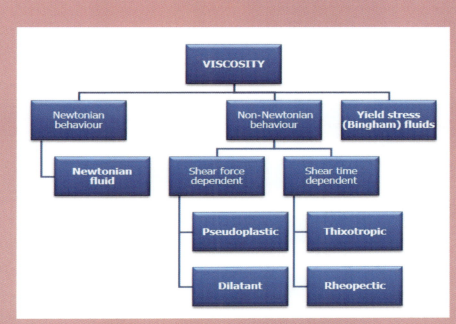

Figure 3. Diagram of the types of viscosity

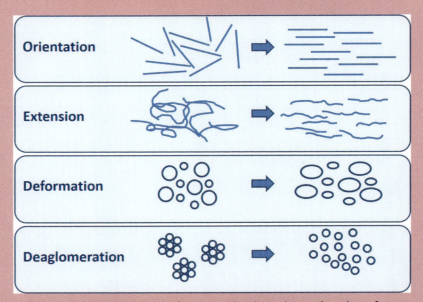

Figure 4. Spatial behavior of macromolecules subjected to shear force

Handbook of hydrocolloids

Table 6. Scheme of viscosity measurement systems

Viscometer	Operation	Pros	Cons
Sensory *"Finger"* 	Dip a finger into the liquid, remove it and let it drip. You can wet one finger of each hand, take them out at the same time and compare which liquid is more viscous.	Cheap Simple Fast Easy to clean	Without units Subjective Without records
Ford Viscosity Cup 	Amount of time until the flow breaks when emptying the cup through a certain diameter hole. Different holes with different diameters are used depending on the viscosity of the fluid.	Cheap Simple Fast Robust Standard DIN	Units (s) that are used in a equation to obtain cSt or cP. Unhelpful in non-Newtonian. Useless in Bingham fluids.
Bostwick consistometer 	Distance covered by a paste during a given time when released in a inclined plane.	Economic Simple Fast Robust ASTM standard	For highly viscous products only.
Falling ball 	Time needed for the ball to go through 100 mm in a tube with an inclination of 10 °.	Economic Simple Fast Standard DIN and ISO	Unhelpful in non-Newtonian. Useless in fluids with initial effort threshold. Useless in opaque liquids.
Rotational	Measuring of the torque of the disc submerged in the solution. Disk size and speed may vary.	Versatile for many types of fluid Data logging For both Newtonian and non-Newtonian fluids	Expensive Delicate Precises careful cleaning of probes and discs.

Hydrocolloids and texture

TEXTURE

PSEUDOPLASTIC?

The word "pseudoplastic" should no longer be used. It's better to use "shear thinning".

However, due to the difficulty of translating this term in some languages, the term "pseudoplastic" continues to be used in some countries.

Rheology of solids

The compounds that do not flow are considered solids, although sometimes, they are so soft that they seem to be highly viscous fluids. Viscosity concepts are not applicable to solids, since the deformation caused by the applied force does not make them flow, but distorts them or breaks them. In these cases, it's necessary to talk about a relationship between the applied compression force and the produced deformation.

To study solid rheology, different equipment and complex and specific instruments are used depending on the type of food being studied. For generic studies, or just to evaluate hydrocolloids, penetrometers or universal texturometers are used.

This basic equipment applies a unidirectional and perpendicular compression force to the surface of the product to be tested. The device measures and records the resistance that opposes the analyzed sample to the applied force.

- Static penetrometer (figure 5): is the simplest device to measure the consistency of a solid. It consists of a stem that can be cone-shaped or flat (usually 1 cm² surface) on which pressure is applied and is registered by a manometer or visually. It measures the maximum compression that the solid supports prior to deformation or breaking or the distance that the probe penetrates on the solid.
The result is often expressed as "gel strength" or "hardness of the gel" and the units used are g/cm².

- Universal texturometer (figure 6): is a more sophisticated equipment that provides more information and, in addition to the consistency or initial penetration, it can also measure other parameters, such as the adhesion and elasticity. For the measurement, a maximum force should be established to implement a rate of descent of the probe, the range of distance that will perform the test, and if deemed suitable, you can program several cycles of measure on the same sample. The result is shown in the form of graphics, and numerical data are recoded.

To better relate the rheological study carried out with the texture in the act of eating or chewing food, it is necessary to be familiarized with the concept "TPA" (Texture Profile Analysis) consisting of consecutively compressing the sample two times, recreating the sensation of chewing in mouth Figure 7.

Different data can be obtained from this type of graph results. Some of them are summarized in table 7.

The rheological properties of some hydrocolloids are measured with lab equipment especially developed for them following special protocols and the result is expressed in specific units (table 8).

Table 7. Interpretation of the TPA graph

Parameter	Definition	Unit
Hardness	Force required to reach a preset deformation	Grams
Elasticity	Relation between the height of the sample at the beginning of the second compression point and the initial height	Adimensional (<1)
Cohesiveness	Relation between the areas under the second and first curve	Adimensional (<1)
Fracturability	The height of the first significant break during the first compression	Grams
Adhesiveness	Negative area below the line profile that represents the work necessary to remove the plunger from the sample base	Grams × mm
Gumminess	Hardness × cohesiveness	Grams
Chewiness (solids)	Hardness × cohesiveness × elasticity	Grams

Table 8. Texture units specific for certain ingredients or additives

Additive	Texturometer	Unit	Definition
Agar	Japanese Agar Gel Tester	Kg/m²	Measure at 20°C on a gel made with 1.5%
Pectin	Ridgelimeter	°SAG	Grams of sucrose that, in an aqueous solution of 65 °Brix and a pH value of 3.2, are gelled by one gram of pectin, obtaining a certain consistency gel.
Gelatin	Bloom gelometer	°Bloom	Required compression (in g) to lower a gelatin gel 4 mm to 6.67%, solubilized at 60°C and cooled 17 hours at 10°C.

Hydrocolloids and texture

TEXTURE

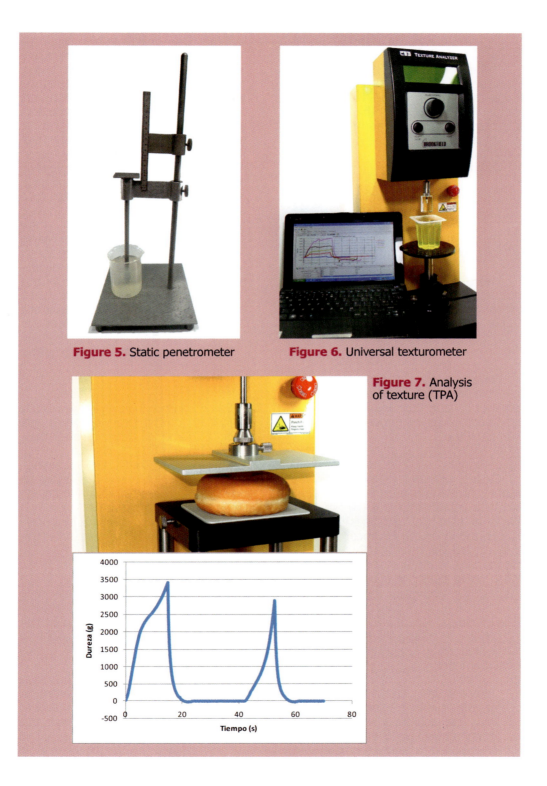

Figure 5. Static penetrometer

Figure 6. Universal texturometer

Figure 7. Analysis of texture (TPA)

Interpretation of measurement data

Properly measuring the rheological properties of a fluid or a solid is not a simple task. To provide reliable and comparable results with other sources it is necessary to control and specify the conditions used when carrying out the tests.

Claims as:

- This hydrocolloid gives a 3500 cPs viscosity.

- This hydrocolloid has gel strength of 1200 g/cm^2.

These are not considered to have interesting meaning unless they are accompanied by complementary information.

Factors that can affect a measurement

Following are a number of factors that can affect the result of the measurements if vary. Consequently, the test conditions must be properly indicated with each test.

Concentration

Clearly, the concentration is an important factor and must be mentioned with the result. Both the viscosity and hardness of a gel increase when the concentration of the hydrocolloid is increased, but not in a directly proportional manner. A double concentration of hydrocolloid does not imply double viscosity or gel strength.

Temperature

Different temperatures must be considered when preparing rheological measurements:

- Temperature used to prepare the samples: some hydrocolloids are soluble in cold or room temperature, while others need to be exposed to higher temperatures. Even the cold soluble products improve their rheological properties if they solubilize after being warmed. The heating temperature should ensure proper dissolution of the hydrocolloid. Thus, heating the sample at 70°C ensures correct solubilization for carrageenan but not for agar or locust bean gum. The solution should also be kept in hot for the proper amount of time to ensure that all the hydrocolloid has been solubilized.

- Temperature used when performing the test: generally, except for some modified cellulose, hydrocolloids increase the viscosity and the gel strength at low temperatures and decrease when temperature increases. Both temperatures of the sample and room temperature need to be controlled separately during testing.

pH

Some hydrocolloids hydrolyze in an acidic medium and lose viscosity or gel strength. Some of them quicker than others.

Some hydrocolloids, such as alginate, become insoluble and precipitate at acidic pH and cease to provide viscosity, or lose their gelling capacity.

On the other hand, some hydrocolloids need very low pH to gel, as it is the case of HM pectin.

Presence of ions

Anionic hydrocolloids especially interact with cations present in the medium. Sometimes, if the concentration of cations is high, you can delay or avoid its solubility, as in the case of CMC, gellan gum or alginate.

Some hydrocolloids, especially the gelling agents, need specific cations to be able to form a gel, and its presence, absence or concentration influences the measurement result.

Hydrocolloids and texture

- K^+: Kappa carrageenan
- Ca^{2+}: alginate, iota carrageenan iota, gellan gum, LM pectin

For this reason, the number of cations must be controlled, or, usually, deionized water must be used for solubilizing hydrocolloids.

The measurement parameters

The viscosity of liquids may vary depending on the shear force and time of application of the same effort. Therefore, those parameters must be specified with the results of a viscosity test:

- The type of probe (spindle) used. Usually they are standard and described by a number.
- The spindle speed (in rpm).
- The time when the measurement was taken, since, like in the case of thixotropic, the behavior of the viscosity will decrease over time.

In the case of gels, it should be indicated the type of probe and its measurements, the speed test, the distance of penetration of the probe and the graphic if a universal texturometer is used.

Standing time

Especially hydrocolloids that solubilize in cold water or at room temperature, require some rest period for achieving total hydration. The viscosity of a recently prepared sample will be slightly lower than the same sample after 24 hours, due to a greater extend on solubilization.

Measurement container

It is also important to standardize the container used to measure the sample because the diameter of it can affect the result of measurement.

In the case of measuring viscosity, the different layers of liquid rotate concentrically, and each one drags the adjoining layer. When the effect arrives at the walls of the container, there will be a braking force that will stop the rotation. If the container is large enough, this effect will be negligible.

In the case of a gel, when the probe presses the surface and the gel tends to flatten, deformation moves to the sides, trying to compensate for the decrease in height with an increase in lateral size. If the container is small, walls impede this deformation, and the mass of the sample offers greater resistance to the probe, obtaining a result of hardness greater than if it had been measured in a larger container.

Examples of technical data sheets of hydrocolloids

It is very risky to accept that the viscosity that will impart a hydrocolloid solution can be summarized in a single data, for example, "4000 cP".

To correctly inform about the rheological aspects, you need to receive much more information, besides the numerical value given by the equipment, as it has been shown.

Below are data appearing in real technical data sheets (TDS) of hydrocolloids marketed for food industry.

Viscosity

Imagine that a company is using a thickener and decides to change it for another. Its only requirement is that the new thickener must provide the same viscosity as the previous one. If they analyze correctly the three TDS received as an alternative, they will see that there is not enough data to decide and choose one of them.

TDS 1	METHYLCELLULOSE
Viscosity 2% in water at 20°C	3500-5600 cPs 4000 rated

The TDS 1 only indicates that 2% of this gum, dissolved in water and measured at 20ºC reaches an average viscosity of 4000 cPs

TDS 2	GUAR GUM
Viscosity:	
after 2h	Minimum 4500 cP
after 24h	More than 5000-5500 cPs

Average viscosity of a 1% solution at 25 ° C. Using a Brookfield RTV viscometer at 20 rpm with spindle 4.

In this case the TDS 2 informs that the viscosity increases if the solution is left to solubilize for enough time.

The concentration is less than in the previous case. You can't compare both TDS. The temperature also changes, but this small difference will not affect the measurement.

More importantly, it indicates the type of instrument used in the measurement, the spindle and the rotational speed. Both spindle and speed are very important data to assess the viscosity in pseudoplastic fluids.

It does not indicate at what time the measure has been determined, so it does not provide information in the case of thixotropic fluid.

TDS 3	XANTHAN GUM
Viscosity	1200 - 1700 cPs

Average viscosity 1% solution. Spindle 3 at 60 rpm

The concentration used matches the previous one. There is no indication of temperature measurement.

The spindle and rotational speed are different from the previous test. There is no possibility of comparison with none of the above.

It does not indicate at what time the measurement has been taken, so it does not provide information in the case of thixotropic fluid

Gelling

Something similar happens if TDS for gelling agents are compared.

In this case, for instance, about carrageenan.

TDS 4	CARRAGEENAN
Gel strength	1000 ± 150 g
Measured at 1.5% at 10°C	
INGREDIENTS:	
Carrageenan (E-407), potassium chloride (E-508) and dextrose to standardize.	

The TDS 4 indicates that the KCl and dextrose have been used to standardize the carrageenan. K+ ions are necessary for the proper functioning of the kappa carrageenan, but there is no indication of how much KCl has been added. If more K+ ions are added to water, the gel strength will increase, as well as the syneresis.

With this information it's impossible to know up to what extent the added K+ ions have influenced the gel strength.

However, the information does not indicate the type of probe (cylindrical, conical, surface...) and conditions used when performing the test.

TDS 5	CARRAGEENAN
Gel strength	650 - 1050 g
Measured at 1.5% at 10°C	
INGREDIENTS:	
Carrageenan, dextrose	

An analysis also appears in this product (per 100g of product):

Sodium	1800 mg
Potassium	15,500 mg
Calcium	100 mg
Magnesium	20 mg

Hydrocolloids and texture

In this other TDS, the measuring temperature and the concentration of the product coincide with the previous.

There is no declaration about added potassium salts that could increase the strength of carrageenan, but the analysis accompanying the TDS 5 shows a high amount of K+. Everything seems to indicate that it is also a kappa carrageenan.

TDS 5 and TDS 6 belong to the same carrageenan producer. It is interesting to compare the amount of potassium and calcium among both samples of carrageenan.

As in the former TDS, the information does not indicate the type of probe (cylindrical, conical, surface...) and conditions used when performing the test.

TDS 6	CARRAGEENAN
Gel strength	145 - 195 g
Measured at 0.21% in milk	
INGREDIENTS:	
Carrageenan, dextrose	
An analysis also appears in this product (per 100 g of product):	
Sodium	1400 mg
Potassium	6600 mg
Calcium	1600 mg
Magnesium	20 mg

The concentration used in this case is much lower, so it is not possible to compare the hardness of the product. The solution has been made with milk instead of water (probably to take advantage of the synergy with ions of Ca^{2+}, which suggests that it is an iota carrageenan).

The measurement temperature is not indicated.

In the list of ingredients, potassium and calcium are not shown to be added.

Compared with the previous one, the amount of potassium and calcium appear to confirm that, in the case of TDS 5 it is a kappa carrageenan and needs potassium ions, and in this case, it is an iota carrageenan needing calcium ions.

The information does not indicate the type of probe (cylindrical, conical, surface...) and conditions used when performing the test.

With the information available from the TDS for the gelling agents, it will not be easy to know what will happen when replacing one with another. Between the two carrageenans supposed to be kappa type, the amount of added potassium will have much influence on gel strength and syneresis. In any case, it is arguable that the third, supposed to be an iota type, will form a gel whose features will be very different from the others.

Conclusion

In the majority of cases, in order to be completely sure, comparative tests should be carried out in our own laboratory or in an external one, as long as the conditions that the samples are exposed to are the same.

2 INTRODUCTION TO HYDROCOLLOIDS

Introduction

Hydrocolloids are macromolecules that have a certainly affinity with water. They are used in food processing precisely for their ability to hydrate themselves and their high solubility in water, modifying its properties to flow. In some cases, hydrocolloids can draw out the flowing of water due to their tendency to create the solid substance, with rigid properties and a variable consistency, that we call gel.

Although hydrocolloids are sometimes referred as "gums", most of them already have the word "gum" in their name. For instance: guar gum or xanthan gum.

When it comes to food technology, hydrocolloids are generally introduced as additives and have their own identification code, which is composed of three or four numbers proceeded by the letter "E". When this manual was written, all hydrocolloids began with the number 4 (not including the modified starches and pullulan), even though other substances, like other stabilizers, phosphates, emulsifiers or polyols, also began with number 4.

Other ingredients, not considered additives, are also hydrocolloids, for example proteins and starches. This manual only focuses on those considered additives, except modified starches.

Most of hydrocolloids are obtained from plants or microorganisms and, in general, they are considered soluble fibers (Table 1). Some of them are chemically modified to enhance or change its function.

According to the hydrocolloid/water ratio and depending on the characteristics of every hydrocolloid, solutions can be of type "sol" (liquids with more or less viscosity) or type "gel" (solids more or less rigid and consistent).

The characteristic of sol or gel depends mostly on temperature, besides the concentration and characteristics of the hydrocolloid. Therefore, hydrocolloids present different behaviors due to:

- Temperature:
  o They can be soluble or insoluble at room temperature.
  o When heated, they can become soluble or insoluble.
  o When heated, the viscosity decreases and recovers when cooled.
  o When heated, viscosity increases and decreases when cooled.
  o It can make a gel when cooled.
  o The gels can either be thermoreversible or thermostable.

- Addition of ions, sugars or acids:
  o Some can make a gel in the presence of divalent cations.
  o Some can make a gel in acidic solutions and there is a high concentration on soluble solids in the medium.
  o Some can become insoluble in acidic solutions.
  o Some can decrease solubility in water in the presence of divalent cations.

Table 2 summarizes these behaviors.

Chemical characteristics

Hydrocolloids are mainly formed by carbohydrates (neutral sugars or their acid or oxidized forms). These form linear or branched chains. Hydrocolloids not only have the basic skeleton or structure of sugars; some can include a protein fraction or chemical groups present in acidic sugars through esterification (Table 3).

Introduction to hydrocolloids

Table 1. Origin of hydrocolloids (photo of microorganisms courtesy of Cristina Madrid)

Origin	Naturals	Chemically modified
Exudates	Arabian Tragacanth Karaya	
Seeds	Locust bean gum Guar gum Tara Cassia	
Algae	Alginate Agar Carrageenan	Propylene glycol alginate
Fruits	HM pectin	LM pectins Amidated pectins
Cereals	Starches	Modified starches
Tubers	Konjac Starches	Modified starches
Cellulose	MCC	CMC MC HPMC
Microorganisms	Xanthan gum Gellan Pullulan	Low acyl gellan

Handbook of hydrocolloids

Table 2. Behavior of hydrocolloids in solution

Cold	When heated	When cooled	Re-heated	Examples
Medium/High	Low viscosity	High viscosity	Low viscosity	Guar gum, xanthan gum, carboxy methyl cellulose, lambda carrageenan
Insoluble	Low viscosity	High viscosity	Low viscosity	Locust bean gum
Insoluble	No viscosity	Gelifies	No viscosity	Carrageenan, agar
Medium viscosity	Viscosity increases	Medium viscosity	Viscosity increases	Hydroxypropyl methyl cellulose
Medium viscosity	Gel	Medium viscosity	Gel	Methyl cellulose
Gelifies with Ca^{++}	Gelifies with Ca^{++}	Gelifies with Ca^{++}	Gelifies with Ca^{++}	Alginate
Solubility depends on Ca^{++}	Low viscosity	Gel with Ca^{++}	Gel with Ca^{++}	LM pectin, Gellan
Partially soluble	Medium viscosity	Gel if pH↓ y SS↑	Gel if pH↓ y SS↑	HM pectin

Introduction to hydrocolloids

Table 3. Chemical characteristics of hydrocolloids

E number	Hydrocolloid	Sugars that form it	Other groups
E-401	Alginate	Guluronic acid Mannuronic acid	
E-406	Agar	**Agarose:** galactose, anhidrogalactose **Agaropectin:** galactose, anhidrogalactose, guluronic acid	Pyruvate Sulfate
E-407	Carrageenan	Galactose, anhidrogalactose	Sulfate groups
E-410	Locust bean gum	Mannose, galactose	
E-412	Guar gum	Mannose, galactose	
E-413	Tragacanth gum	**Bassorin:** xylose, fucose, galacturonic acid, galactose **Tragacanthin:** arabinose, galactose, fucose, xylose, galacturonic acid, rhamnose Remains of protein Starch	Methoxyl
E-414	Gum arabic	Galactose Arabinose Rhamnose Glucuronic acid Proteins	
E-415	Xanthan gum	Glucose Mannose Glucuronic acid	Pyruvate Acetyl
E-416	Karaya gum	Galactose Rhamnose Galacturonic acid Glucuronic acid	Acetyl
E-417	Tara gum	Mannose Galactose	
E-418	Gellan	Glucose Rhamnose Glucuronic acid	Acetyl Glyceryl
E-425	Konjac	Mannose Glucose	Acetyl
E-427	Cassia gum	Mannose, galactose	
E-440	Pectin	Galacturonic acid Rhamnose	Methyl
E-460	Microcrystalline cellulose	Glucose	
E-461	Methyl cellulose	Glucose	Methyl
E-464	Hydroxypropyl methyl cellulose	Glucose	Hydroxypropyl Methyl
E-466	Carboxy methyl cellulose	Glucose	Carboxy Methyl
E-1204	Pullulan	Glucose	

Handbook of hydrocolloids

Some hydrocolloids are composed of a single type of sugar, (known as homopolysaccharides), for example, cellulose derivatives. Others, however, combine different sugars (known as heteropolysaccharides), such as alginate or guar gum.

Usually, homopolysaccharides tend to form linear spatial structures (cellulose) or helical (amylose). Meanwhile, heteropolysaccharides tend to generate spatial structures in zigzag (alginate) or act as branched chains (guar gum).

In the case of the heteropolysaccharides, the ratio and distribution of the different sugars in the macromolecule can influence the solubility (as in the case of galactomannans) or its interaction with some ions present in the solution (as it is in the case of alginate).

Due to the fact that these macromolecules are basically formed by sugars, and they can be digested in the upper intestinal tract, they can also be considered as soluble fibers.

Depending on the presence of acid carboxylic groups and other different substitutes, hydrocolloids belong to one of two groups, ionic (mainly anionic due to their negative charge) or non-ionic (Table 4). Only some hydrocolloids, not considered additives but ingredients, are cationic, as in the case of gelatin from pigskin and chitosan.

Generally, variations in pH and the presence of free cations have a greater influence on ionic hydrocolloids than non-ionic. Depending on the pKa and pH of the solution, the carboxylic group of sugars is more or less ionized (Figure 1). Some hydrocolloids, as alginate, become insoluble in its acid form and precipitate.

Cations present in the solution will interact with the negative charges of the acid groups (Table 5). Because of this, the majority of the anionic hydrocolloids portray different behaviors depending on the amount and type of cations present in the solution. Sometimes they become insoluble, but sometimes, they form three-dimensional networks. This effect can be beneficial, for instance, when we need alginate to react immediately after being in contact with free calcium ions.

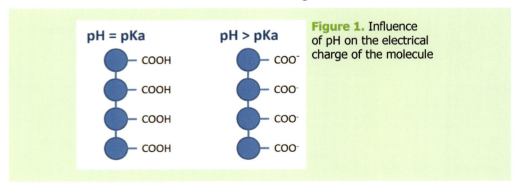

Figure 1. Influence of pH on the electrical charge of the molecule

Table 4. Classification of hydrocolloids according to their electrical charges

Ionic		Non-ionic	
Alginate	CMC	Guar gum	MC
Carrageenan	Gellan	LBG	HPMC
Agar (Agaropectin)	Xanthan gum	Tara	Konjac
Tragacanth	Pectin	Microcrystalline cellulose	Agar (agarose)
Karaya	Arabic		

Introduction to hydrocolloids

When negatively charged groups are accumulated in some areas of the hydrocolloid, it can interact with proteins that have positively charged amino acids. For example, this can happen when the kappa carrageenan interacts with casein to form a weak three-dimensional network that helps keep suspended cocoa particles in milk beverages.

The structure and spatial conformation of molecules is important to understand some of the qualities of the hydrocolloids, such as their ability to solubilize in cold conditions (usually branched macromolecules or with ionized groups) or, on the contrary, need a previous warm-up (usually linear macromolecules and/or non-ionized groups).

The spatial structure of hydrocolloids can give an idea of its possible behavior once solubilized in an aqueous phase:

- Thickeners: usually linear chains.

- Gelling agents: usually branched, with anionic groups or chains with linear areas that, with the help of hydrogen bonds, are able to interact with other linear zones of themselves or other hydrocolloids, enhancing different synergies among them.

Hydration and solubilization of hydrocolloids

Hydrocolloids, as its name suggests - hydro-, need water to solubilize and be able to execute their technological function.

The molecules found in non-ionic linear hydrocolloids (or with few ramifications), easily order themselves one next to the other, as if they were being "packaged", thanks to the hydrogen bonds present. To separate these molecules and to solubilize them, an external energy source is required, which is mainly heat. Because of this, these types of hydrocolloids are insoluble in cold water and become soluble when exposed to heat.

Branched molecules prevent the formation of this packaging favoring the entry of water between molecules and, therefore, the solubilization of the hydrocolloid. The same happens when the hydrocolloid is anionic, and the negative charges repel each other, separating the molecules. These types of hydrocolloids are also soluble in cold water.

The presence of positive ions in aqueous solution (for instance pickles, hard water or milk) can hinder the solubilization of anionic hydrocolloids. In fact, the cations present in the solution can block the repulsion charges and avoid the creation of bridges between them. In these cases, it may be necessary to heat the solution or subject it

Table 5. Interactions due to electrical charges

Anionic hydrocolloids and proton interactions		
E-401	Alginate	Ca^{2+}
E-407	Kappa carrageenan	K^+
E-407	Iota carrageenan	Ca^{2+}
E-418	Gellan	Ca^{2+}
E-440	HM pectin	Ácidos pH↓
E-440	LM pectin	Ca^{2+}
E-466	Carboxy methyl cellulose	Ca^{2+}

Handbook of hydrocolloids

to an intensive mechanical work, such as homogenization. It is often said that, in these cases, the "activation" of the hydrocolloids is needed, meaning it is necessary to increase the shear ratio.

Preparation of hydrocolloids solutions

To correctly disperse and dissolve hydrocolloids is often a challenge. The addition of a fine powder on a watery liquid produces lumps, and when the addition of powder is done not carefully, the external part of the lump of powder becomes wet, sealing their surface and preventing water from penetrating.

The situation worsens when this powdery product is too eager for water, and it creates a thick viscous layer, making the dispersion of the powder difficult. This tends to happen in hydrocolloids that are soluble in low temperatures.

An incorrect dispersion and dissolution of hydrocolloids causes:

- Unpleasant lumpy aspect.

- Strange and unpleasant mouthfeel.

- Loss of the thickener/gelling power due to the fact that this part of the product is not soluble.

- Possible clogging of filters, nozzles or valves.

Therefore, dispersion must be done without lumping and with a proper, careful, separation of the powdery particles. For this reason, it is advisable to:

- Previously disperse the hydrocolloid in an environment where it is insoluble (oil, alcohol, acidic solutions, syrups or brine).

- When available on the market, use granulated hydrocolloids, which are not so powdery and have larger particles.

- Make a premix with other powdered products that are easily dispersible in water, such as salt or sugar. The mixing ratio must always be favorable to the dispersible product, at least to a ratio 1:5.

- Add the hydrocolloid slowly into the liquid, using a mechanism that mixes and shakes the liquid.

- Use equipment specially designed for these purposes where the liquid is forced to cycle through, dragging and slowly adding the hydrocolloid (figure 2).

Specifications

Like any other ingredient or additive, commercial hydrocolloids must be accompanied by a proper data sheet that provides as much information as possible. The amount of information provided by the manufacturer/distributor of the hydrocolloid varies with each company (Table 6).

In general, it is possible to distinguish the basic specifications all manufacturers must show and the complementary specifications that would be very well received by the customer.

The quality and purity of the hydrocolloids specifications, as well as for other additives, are collected by an EU regulation. At the time of writing the manual, the active regulation is the *Commission Regulation (EU) No 231/2012 of 9 March 2012 laying down specifications for food additives listed in Annexes II and III to Regulation (EC) No 1333/2008 of the European Parliament and of the Council* (and its subsequent amendments).

Technological functions

The first thing that comes to mind when thinking about hydrocolloids is their ability to, in some way, immobilize water. For this reason, as mentioned earlier, the main function of hydrocolloids is to act as thickeners or gelling

Introduction to hydrocolloids

agents that modify the rheology of foodstuff, either increasing viscosity or creating a more or less rigid structure, called gel.

Thickeners

Thickeners increase the viscosity of the liquid when they interact and solubilize in it. Hydrated macromolecules interact with each other through forces of cohesion, moving and colliding, complicating the capacity of liquid flowing.

Usually, when fluid is put in motion by shaking, pumping or draining by

Figure 2. In-line mixer of powdery products. Courtesy of Vakkimsa.

Table 6. Technical specifications

Basic specifications	Complementary specifications
Name	Characteristics of dispersion and hydration
Chemical composition	Formation of turbid or clear solutions
pH in solution	Freezing / defreezing stability
Particle size	Stability or tolerance to acidity, temperature, alcohol or salt
Viscosity or gel strength	Possible interactions with other components
Presence of allergens	Possible synergies
Heavy metal limit	Aspect (colour, presence of dark particles, ...)
Microbiology characteristics	Presence of active vegetables enzymes
Safety data sheet	

gravity, the apparent viscosity tends to decrease. This is due to macromolecules breaking down, or orienting themselves in the direction of the flow, decreasing the amount of friction between them and, therefore, the viscosity of the liquid.

Gelling agents

When it comes to gelling agents, water remains trapped and immobilized in a three-dimensional network of molecules that is structured by (figure 3):

- The occurrence of crystalline zones caused by packaging due to hydrogen bonds. It occurs in non-ionic macromolecules with short or no branches. The packaging can occur between molecules of the same or two different hydrocolloids (for instance in the case of xanthan and locust bean gum).

- The creation of bonds caused by divalent cations such as calcium (such as in the case of the alginate, gellan gum or LM pectin).

- The presence of hydrophobic interactions due to the disappearance of the anions when pH is around its pKa (for instance in the case of HM pectin) or to an increase in temperature (methyl cellulose).

- The formation of covalent bonds as, for example, disulphide bridges (-S-S-) due to heating or the presence of oxidizing agents. This phenomenon occurs in proteins that form gels when exposed to heat, but these are not the objective of the study presented in this manual.

In both thickeners and gelling agents, the size of molecule – or molecular weight – is critical in order to execute their functions with greater or lesser intensity. Longer molecules have a greater ability to create three-dimensional networks or, when moving through a liquid, they interfere more intensely in the flow of fluid. Consequently, the longer the length of the string, or the higher the molecular weight of the molecule is, more effective is the thickener or gelling agent. Accordingly, any mechanism that breaks or cuts the molecules (mechanical, acid or enzymatic hydrolysis...) causes a drastic decrease on the expected viscosity or the strength of the gel.

By imaging in a fluid a linear molecule of a particular size, which can rotate freely in all directions, it may rub or collide with other similar molecules entering its action site. This space would correspond to the volume of a sphere,

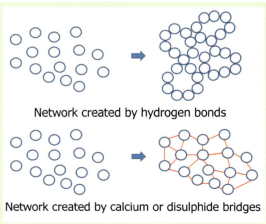

Figure 3.
Organization of the molecules to make a gel

Introduction to hydrocolloids

and the distance from the center to the circumference being represented as its radious (r).

Figure 4 shows the volume of space in which this molecule can interact.

Contact with other molecules that penetrate in the action site will result in the increase of viscosity or the possibility of interaction and formation of a gel.

If hydrolysis decreases the size of the molecule by half, the volume of the occupied space would also decrease to 1/8 of its original volume, so the possibility of interaction with other molecules will be much smaller, and therefore the viscosity or the gel strength will decrease considerably.

Since hydrocolloids have the ability to increase viscosity or form a gel, they have non less important secondary technological abilities. Sometimes, the choice of one or another hydrocolloid depends more on these secondary features than on its own ability to provide viscosity or form a gel.

The main secondary technological functions are:

- Suspension of insoluble particles dispersed in the medium: some hydrocolloids create a very weak three-dimensional network, without gelling or significantly increasing the viscosity, but hindering the fall or decanting of insoluble particles. Some hydrocolloids interact with proteins, due to their formula and electrical charges, forming a weak network. For example, in the case of the kappa carrageenan.
Others such as gellan gum interact with ions, mainly calcium, to form this network. Others, such as xanthan gum, are branched and form the network interacting with themselves. Some structures, like gellan gum or xanthan, are such very weak gels that they cannot manage to immobilize water, and are characterized as "fluid gels".

- Stabilization of emulsions and foams: according to Stoke's law (figure 5) the decantation rate of a sphere moving through a viscous fluid is inversely proportional to the viscosity of the fluid. Therefore, any increase of viscosity in the continuous phase of the emulsion will decrease the speed of separation of the dispersed phase. This is relevant to those hydrocolloids that are able to increase the viscosity of the gel. On the other hand, the possibility of placing macromolecules that are not soluble in the external interface of an emulsion, would delay the interaction of the particles in this dispersed phase

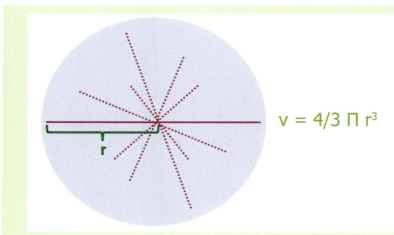

Figure 4. Volume taken up by a molecule in a fluid

$$v = 4/3 \, \Pi \, r^3$$

and would increase the diameter of the sphere. An example of this is microcrystalline cellulose, or MCC. Stokes' law also identifies the importance of maintaining the radius of the particles of the dispersed phase as small as possible, since the speed of decantation of the sphere is directly proportional to the radius. However, hydrocolloids play no role in this.

- Film formation: when a thin sheet of hydrated hydrocolloid dries up, it forms a film. These films can be transparent or translucent, with a greater or less ability to tolerate tensions and tractions depending on the hydrocolloid. In some occasions, this film can also protect against oxidation, since the permeability to oxygen is low.

- Ability to act as a total or partial fat substitute (fat replacer): an increase of the viscosity or the formation of a soft gel can imitate the creaminess effect and texture of fats.

- Encapsulation of flavors.

- Binder

The following chapters will present different hydrocolloids used in food technology and will describe their technical features and their most common uses.

$$V_s = 2r^2 g (\rho_p - \rho_f) / 9\eta$$

Vs	Separation velocity
r	Radius of the sphere
g	Force of gravity
ρ_p	Density of the liquid
ρ_f	Density of the sphere
η	Viscosity

Figure 5. Stokes' law

3 E-400 ALGINATES

E-400 Alginic acid (CAS Number: 9005-32-7)

E-401 Sodium alginate (CAS Number: 9005-38-3)

E-402 Potassium alginate (CAS Number: 9005-36-1)

E-403 Ammonium alginate (CAS Number: 9005-34-9)

E-404 Calcium alginate (CAS Number: 9005-35-0)

E-405 Propylene glycol alginate (CAS Number: 9005-37-2)

Origin

Alginates are extracted from brown algae in the Phaeophyceae class, where alginic acid and salts like sodium, potassium, ammonium and calcium can be found in different proportions.

PGA or propylene glycol alginate is a chemical derivative of alginate obtained from esterification of propylene oxide with neutralized alginic acid (Figure 1).

> *i* From all salts of alginic acid which are currently supported in legislation, sodium alginate is the most commonly used in food industry

Composition and chemical structure

Alginates are lineal polysaccharides formed by sequences of guluronic acid (G) and mannuronic acid (M). They are connected by β (1-4) links that are in different proportions according to the raw material of extraction. Depending on their proportions in the molecule, they can lead to either high-M or high-G alginates (Figure 2 and Figure 5), with different gelling characteristics.

- **High-M alginates:** most of the chains that form them are lineal (Figure 3). They have the purpose to give a spatial configuration, since the mannuronic is not reactive with calcium. They make weak, elastic, flexible gels and with low syneresis.

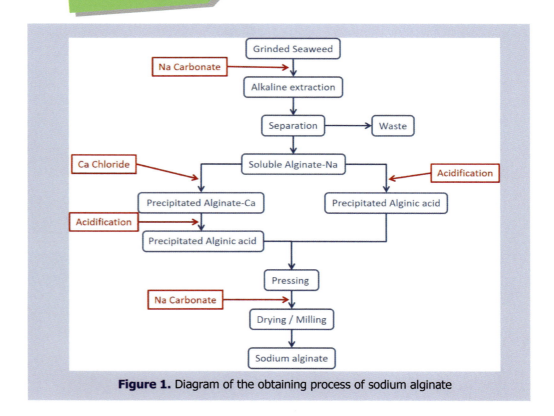

Figure 1. Diagram of the obtaining process of sodium alginate

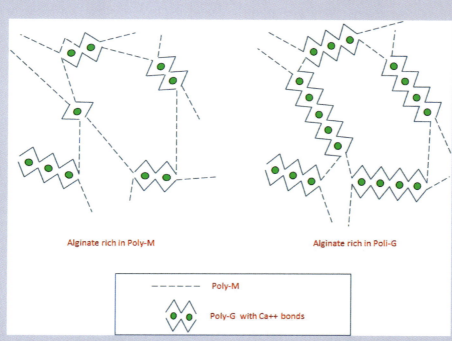

Figure 2. Molecular structure of poly-M and poly-G alginates

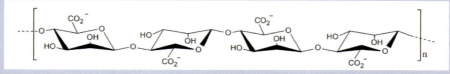

Figure 3. Mannuronic chains (poly-M blocks)

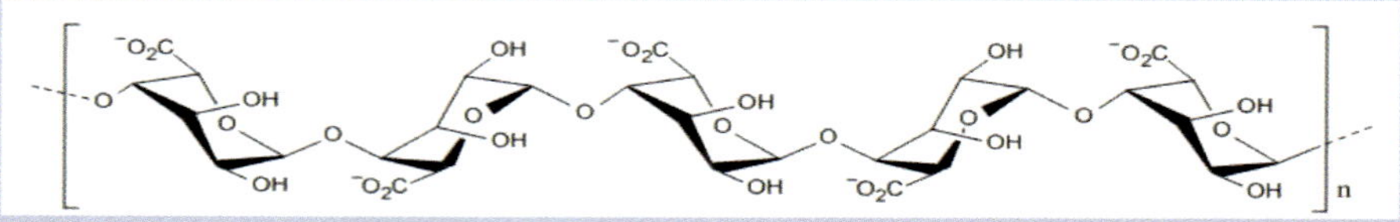

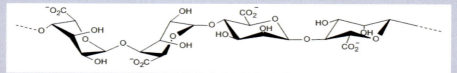

Figure 4. Guluronic chains (poly-G blocks)

Figure 5. Union of poly-M and poly-G blocks

- **High-G alginates:** they are formed by chains with a zigzag structure (Figure 4), making a better connection with calcium. This makes possible the formation of the "egg box" structure (Figure 7). Gels formed with this kind of alginate are consistent, rigid, brittle and more prone to syneresis.

Properties of solutions

Solubility and solutions preparation

Alginic acid and its bivalent and trivalent salts are insoluble and therefore they don't have functionality, unless they are dissolved in an alkaline medium.

If soluble salts are directly added to water they tend to form lumps, so it is important to remember the measurements to ease the correct solution in the medium, specified in chapter 2. In the specific case of alginate, it is important to incorporate the lowest possible air during solubilization, set a defoaming time or apply vacuum to the solution before the gelation (Table 1). The final result of the product can vary significantly as we see in Figure 6, both in appearance and in the volume of the sample.

Factors affecting the properties of solutions

Soluble alginates, such as sodium, potassium, ammonium and propylene glycol act as thixotropic and pseudoplastic thickeners in distilled water or in the absence of calcium ions in the medium.

Alginates are sensitive to mechanical work and have a good behavior in freeze-defreeze cycles. They can slightly increase the viscosity in the presence of calcium in the environment, but a strong heat treatment causes a loss of the same (Table 2).

Functionality

Gelation mechanism

The main feature of alginates is their ability to form gels in the presence

Figure 6. Alginate sample right after applying mechanical work in solubilization (left) and after a resting time to defoam (right)

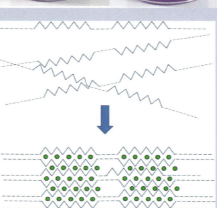

Figure 7. "Egg box" structure

Alginates

E-400

Table 1. Solubility and solutions preparation / +++ (high) ++ (middle) + (low)

Solubility in cold water	Solubilization temperature	Premix	Shear force	Air incorporation	Increase viscosity of medium	Precipitates with alcohol	Ion sequestrant
+++	cold	+++	+++	++	+	+	+++

Table 2. Factors affecting the properties of solutions

Factors	Effects on viscosity	Observations
Molecular weight / polymerization degree	Directly proportional	
Concentration	Directly proportional	
Temperature	Inversely proportional	An increase in temperature causes a decrease in viscosity. Freezing and defreezing cycles do not affect
pH	It is stable over a wide pH range	PGA is soluble at a pH range of 2-7. Starts to saponify in basic solutions
Ionic charge	Small concentrations of Ca^{2+} increase the viscosity	Without the use of a sequestrant, higher concentrations of Ca^{2+} can produce the gelling of the medium
Mechanical work	Pseudoplastic behaviour	It decreases the viscosity as it increases the shear force
Presence of alcohol	It causes an increase of viscosity	In excess, it can cause the precipitation of the alginate molecule

of divalent ions, mainly calcium, with the exception of magnesium.

Its gelation mechanism depends on the presence of calcium ions in the solution (Figure 9). They form bridges between two free carboxyl groups of the gulurónico molecules, creating the structure known as "egg box" (Figure 7). If there are not enough calcium ions in the medium it only produces an increase in the viscosity.

The gel formation occurs very quickly, consequently, in some cases a calcium sequestrant is necessary (phosphates are normally used) to control the release of calcium in the solution and ensure a correct solubilization of alginate and a homogeneous gelling.

Therefore it is very important to know the correct ratio of alginate, calcium salt and sequestrant (Figure 8):

- **Alginate:** a higher concentration will make firmer gels but with strange textures. The M/G ratio indicates the behavior of the gel.

- **Calcium salt:** its concentration and richness in calcium determine the characteristics of the gel. If there is too much calcium in the medium, the gel will be strong but with a grainy texture, and if there is not enough, the gel will be very weak or it will not form. It is also important to know the solubility of the calcium salt (Table 3), since it will determine the gelling speed. Very soluble salts, like calcium chloride, will release the calcium to the medium very quickly, and the gelation will occur immediately when it's in contact with the alginate.

- **Sequestrant:** the dose should be well adjusted. If there is too much concentration in the medium, the formation of the gel will be greatly delayed. If, on the other hand, the amount of sequestrant is too low, calcium-alginate interactions will be faster, and there will be no time for a correct solubilization of the alginate, or a correct mixture with other ingredients. The most commonly used are phosphates, particularly the STTP or pentasodium tripolyphosphate.

Propylene glycol alginate is less reactive to the action of calcium and forms weak gels. Its main function is foaming and foam stabilization, in fact, it is considered more an emulsifier than a hydrocolloid.

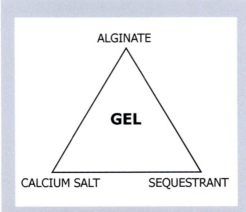

Figure 8. Ratio of alginate, calcium salt and sequestrant

> ⓘ When using calcium chloride as a source of calcium for the gelation (direct external diffusion) it is necessary to wash the formed product with water, since this calcium salt gives bitter taste. However, calcium lactate does not present this problem.

Table 3. Main characteristics of the calcium salts used in alginate gelation

Calcium salt	Molecular weight (g/mol)	% Ca^{2+}	Solubility in water at 20ºC (g/100 ml)	Most frequently used
Calcium chloride Cl$_2$Ca	110,98	36	74,5	Direct external
Calcium lactate 5-Hdrte C$_6$H$_{10}$CaO$_6$·5H$_2$O	308,29	13	6,6	Indirect external Direct external
Anhydrous calcium sulphate CaSO$_4$	136,14	29	0,24	Direct internal
Calcium sulfate 2-Hdrte CaSO$_4$·2H$_2$O	172,17	23	0,24	Direct internal
Calcium carbonate CaCO$_3$	100,08	40	0,0013	Indirect internal
Calcium citrate CA$_3$(C$_6$H$_5$O$_7$)$_2$	498,43	24	0,095	Indirect internal
Calcium Gluconate C$_{12}$H$_{22}$O$_{14}$	430,37	33	3,5	Indirect internal

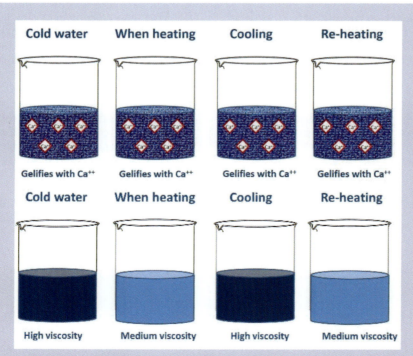

Figure 9. Solubilization and gelation scheme of sodium alginate in a medium with calcium (up) and without calcium (down)

Handbook of hydrocolloids

Form of preparation

Alginate gel is formed with the union of alginate chains with calcium ions. The gelation can be achieved through several methods (Table 4):

- **Direct external diffusion:** it allows the making of structured products in a fast way, almost immediately (Figure 11). It is obtained by dosing an alginate solution with the product to gel over a bath of water with a dissolved calcium salt. Once it is dosed on the bath with the calcium salt, the surface gels immediately and increases as calcium enters the gel (Figure 10). The most used calcium salt in this procedure is calcium chloride, since it is very soluble but leaves a bitter taste that may be unpleasant. Calcium lactate is relatively less soluble but does not have residual flavors.

- **Indirect external diffusion** (also called reverse): follows the same principle as the previous gelling, but in this case, the product contains calcium and it is dosed over a bath with alginate to low concentration, so that the product can penetrate and the entire piece gets covered by the alginate.

- **Direct internal diffusion:** the gelation occurs slowly; it is not immediate like in the previous methods. It allows us to make structured products in a mould with a defined shape. An alginate solution with a sequestrant is prepared and mixed with the product to gel. Then a low solubility calcium salt, such as calcium sulfate, is added, which releases calcium in the solution gradually. The use of this technique of slower gelation makes it possible to use alginate to develop larger structures, in blocks or shapes, which would take a lot of time to gel in an immersion bath (Figure 12). By using the soluble calcium salts used in the direct method, the gel is formed too fast, even before giving the desired form to the product, and the gelled part breaks when kneaded.

- **Indirect internal diffusion:** the system is similar to the direct internal diffusion. It requires a combined system of alginate, sequestrant, acidifier and calcium salt, but an insoluble calcium salt (for instance calcium carbonate) is used instead, which is solubilized when there is a decrease in the pH of the solution and releases calcium in the medium, producing the gelation. This decrease in the pH of the medium must be done gradually and using regulators of pH, such as the glucono delta-lactone (GDL) or a mixture of citric acid and sodium citrate.

Factors affecting gel properties

- **Temperature:** at high temperatures the chains have an excess of

Figure 10. Example of a cross section of sodium alginate gelation with calcium chloride by the direct external diffusion method

Alginates

energy and they can't be aligned, so they can't form the gel. Temperature can be also used as a form of preparation of gels, by heating the alginate solution and adding the calcium to the medium while it is still hot. By doing this, the solution will gel in the desired mould (Figure 13).

- **pH:** a decrease in the pH of the medium increases the gelling efficiency because there are fewer free calcium ions needed to gel, but an excessive decrease causes alginate precipitation.

- **Calcium sources/sequestrant:** present in the product to gel.

- **Sugars:** high concentrations of soluble sugars reduce the gel strength.

Characteristics of gels

Depending on the proportion of mannuronic and guluronic molecules in composition gels made with alginate by any of the diffusion methods, they may have different characteristics mostly related to texture (Table 5).

Synergies or incompatibilities

- Sodium alginate + starches: starches block free carboxyl groups of alginate by hydrogen bonding; consequently, more activation energy is needed to break these bonds and replace them with calcium. This causes the obtained gels to be less elastic.

- Sodium alginate + pectin: they can form thermoreversible gels in acidic

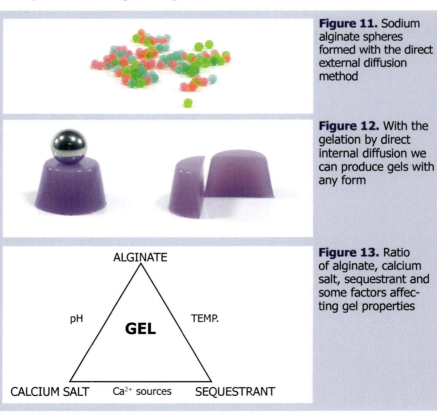

Figure 11. Sodium alginate spheres formed with the direct external diffusion method

Figure 12. With the gelation by direct internal diffusion we can produce gels with any form

Figure 13. Ratio of alginate, calcium salt, sequestrant and some factors affecting gel properties

Handbook of hydrocolloids

Table 4. Forms of preparation of alginate gels

	Type of diffusion		In the product	In the immersion bath	Sequestrant
Broad-casting	External	Direct	Alginate	Calcium salt	In the product (*)
		Indirect	Calcium salt	Alginate	In the bath (*)
	Internal	Direct	Alginate + insoluble calcium salt	---	In the product
		Indirect	Alginate + insoluble calcium salt + pH regulator	---	In the product

(*) If it is necessary

Table 5. Characteristics of gels

Type of alginate	High-M alginates	High-G alginates
Gel type	Weak and elastic	Firm and crisp
Syneresis	No	Yes, after freeze-defreeze
Temp. stability	Thermostable	Thermostable
Synergies	Starch and pectin	Starch and pectin

conditions, without the presence of calcium in the solution.

Examples of applications in food industry

- Gelling agent in pepper or anchovy paste for fillings. Gelation occurs by direct external diffusion, using calcium chloride as a source of calcium.

OLIVES STUFFED WITH PEPPER PASTE

Ingredients: green manzanilla olives, pepper paste (20%) (pepper, stabilizers (**sodium alginate**, guar gum, **calcium chloride**)), salt, acidity regulator (citric acid), antioxidant (L-Ascorbic acid).

- Formation of films, casings, coatings, etc. In this case, it is sought to form a film with alginate which makes the sausage casing, taking advantage of its thermostable properties.

TURKEY AND CHICKEN SAUSAGES

Ingredients: turkey meat (65%), chicken meat (15%), water, salt, soy protein, starch, spices, sugar, humectant (sorbitol), preservative (potassium metabisulphite), antioxidant (ascorbic acid and sodium citrate), gelling agent (**sodium alginate**), stabilizer (guar gum), acidity regulator (acetic acid, lactic acid and **calcium sulfate**), colour (cochineal).

- Ice cream stabilizer.

STRACCIATELLA ICE CREAM

Ingredients: water, sugar, corn and wheat glucose syrup, chocolate shavings (6,5%) (sugar, cocoa paste, cocoa butter, emulsifier (soy lecithin), flavorings), coconut oil, skimmed milk powder (6%), milk derivates, emulsifier (mono- and di-glycerides of fatty acids), stabilizers (**sodium alginate**, locust bean gum), flavorings.

Alginates

E-400

- Stabilizer in instant or cold milk desserts. The product is dissolved in a calcium-rich medium (milk) so we need to use a sequestrant, such as as potassium phosphate, citric acid and citrate sodium to control the gelation. In this case, the alginate is used as a mousse stabilizer.

> **LEMON MOUSSE MIX**
>
> Ingredients: sugar, glucose syrup, vegetable fat, gelatin, acidity regulators (**citric acid, sodium citrate**), lactose, emulsifier (mono and diglycerides of fatty acids and acetic esters of mono and diglycerides of fatty acids), milk proteins, stabilizers (**sodium alginate** and **potassium phosphate**), salt, flavorings and colours.

- Thickener in bakery fillings and sauces.

> **CHOUX PASTRY FILLED WITH COFFEE CREAM**
>
> Ingredients: water, sugar, eggs, flour, milk powder, butter (4,9%), modified starch, cream (2,5%), glucose syrup, non-hydrogenated vegetable fat (palm, palm seed), skimmed milk powder, instant coffee (0,9%), stabilizers (carrageenan, glycerol), gelling agents (**diphosphates, sodium alginate**, agar), salt, emulsifier (citric esters of mono- and diglycerides of fatty acids, lecithin), wheat gluten, preservative (potassium sorbate), flour treatment agent (ascorbic acid).

- The PGA is mostly used as beer foam stabilizer.

> **BEER**
>
> Ingredients: water, barley malt, corn, hops, yeast, stabilizer (**propylene glycol alginate**), antioxidant (ascorbic acid).

🔍 Spherification

Alginate gelation by direct external diffusion is widely described in the bibliography and is commonly used in the food industry. However, chef Ferran Adrià introduced its use in a professional kitchen and popularized the term "spherification" to refer to the technique of creating spheres from this method.

4 E-406 AGAR

E-406 Agar (CAS Number: 9002-18-0)

Other denominations: agar-agar, kanten

Handbook of hydrocolloids

Origin

It is obtained from red algae's cell walls, also called agarophyte algae, in particular of some species of the genus *Gelidium*, *Pterocladia*, *Gelidiella*, *Gracilaria* and *Gracilariopsis* (Figure 1).

Composition and chemical structure

It is a heteropolysaccharide formed by two polysaccharides: agarose (Figure 2) and agaropectin (Figure 3).

The proportion between the two types of molecules depends on the type of algae and the extraction process, but is generally 70% agarose and 30% agaropectin. (Figure 4) Table 1 describes the main differences between the two molecules.

Properties of solutions

Solubility and solutions preparation

Agar is insoluble in cold water. It starts to solubilize in hot water, becoming completely soluble in boiling water.

As with most hydrocolloids, it is recommended to premix agar with other powdered products before adding it to the aqueous medium and apply shear force to disperse it correctly, although it does not tend to form lumps.

It does not incorporate air in the mix process, so it is not necessary to leave a rest time for defoaming. It also does not provide a significant increase in the

Table 1. Differences between molecules of agarose and agaropectin

Molecule	Ionic charge	Gelling capacity	Sulfate content
Agarose	Neutral	Yes	Very low
Agaropectin	Negative charge	No	High

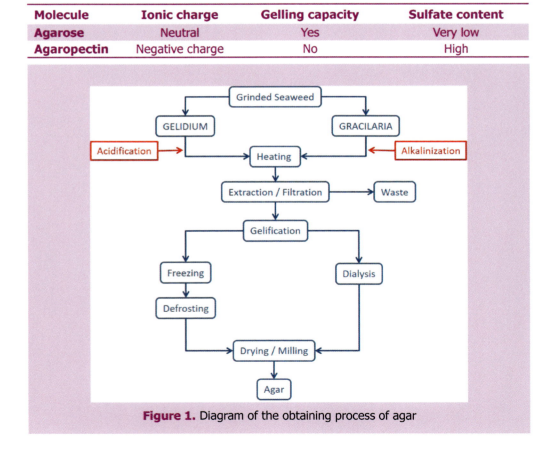

Figure 1. Diagram of the obtaining process of agar

viscosity of the medium and is stable in the presence of high concentrations of alcohol (Table 2).

Factors affecting the properties of solutions

- **Temperature:** it must reach a high enough temperature to completely solubilize agar, it is recommended to exceed 90ºC (Table 3).

- **pH:** in acidic mediuma, a hydrolysis of the agar molecule is produced. If the medium is not neutralized with a base before the solubilization of the agar, it may cause the loss of the gelling ability. In Figure 5 you can see the hardness of gels formed from the same solution of 1% agar in water, which was adjusted to pH 2/5/7,5 and then heated to solubilize the agar. Solutions at pH 5 and 7,5 produce gels with similar hardness. In the sample adjusted at pH 2, hardness is much lower, so it cannot form a gel.

Functionality

Gelation mechanism

After the solubilization of agar at close to boiling temperatures, the gelation does not occur until the temperature is

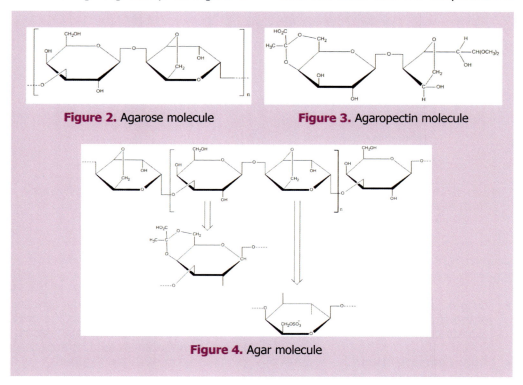

Figure 2. Agarose molecule

Figure 3. Agaropectin molecule

Figure 4. Agar molecule

Table 2. Solubility and solutions preparation / +++ (high) ++ (middle) + (low)

Solubility in cold water	Solubilization temperature	Premix	Shear force	Air incorporation	Increase viscosity of medium	Precipitates with alcohol
Insoluble	+++	+	+	No	No	No

Handbook of hydrocolloids

Table 3. Factors affecting the properties of solutions

Factors	Effects on viscosity	Observations
Molecular weight / polymerization degree	Directly proportional	
Concentration	Does not affect	
Temperature	Does not affect	
pH	It is stable over a wide pH range	
Ionic charge	Does not affect	
Mechanical work	Pseudoplastic behaviour	It decreases the viscosity as it increases the shear force

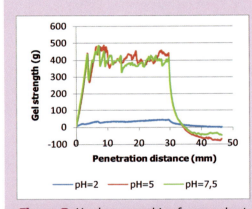

Figure 5. Hardness graphic of agar gels at 1% in relation to the pH of the medium

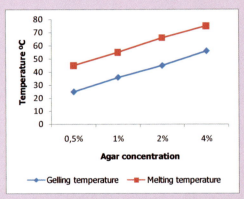

Figure 6. Graphics of different gelling and melting temperatures based on agar concentration

between 25ºC and 45ºC (Figure 7). This temperature range depends on:

- **Origin:** depending on the genus of the algae from which the agar is extracted.

- **Agar concentration:** the hysteresis loop depends on the agar concentration in the medium. Likewise, at higher concentrations of agar, gelling and melting temperatures increase, as shown in Figure 6. Therefore, the temperature increase is directly proportional to the increase of agar concentration.

Once the agar gel is formed, it is possible to melt it (thermoreversible gel) at temperatures between 60-85ºC (could be lower depending on the agar concentration) and keep it liquid until a drop at gelling temperatures. This feature, known as a hysteresis loop, is especially pronounced in the agar (Figure 8).

Characteristics of gels

Agar gels tend to be firm and brittle (Figure 12). An increase in agar concentration means an increase in the gel hardness. Figure 9 shows how, by doubling agar concentration, the hardness of the formed gel increases significantly.

Likewise, adding an alcohol molecule to the medium, in this case ethanol, does not affect negatively the solubilization or gelation of agar but is enhanced in the gel strength (Figure 10).

The highest stability of the agar gel is a pH 5-9. In very acidic or very alkaline solutions there is a hydrolysis of the molecules and a loss of the gel-forming ability. Another factor that can affect the gelation of agar is the ion charge of the medium. High concentrations of NaCl in the medium (around 5%) prevents the correct solubilization of the agar, making it impossible for it to gel. On the contrary, in tests with high concentrations of sugar, no noticeable difference appears in the gelling ability and hardness of the gels obtained, as shown in Figure 11.

Agar gels have syneresis, but this decreases as we increase the agar concentration and, therefore, increases the gel strength. They are translucent and whitish. In the presence of high concentrations of sugar, glucose or glycerol, transparency and brightness of the gels are increased (Table 4).

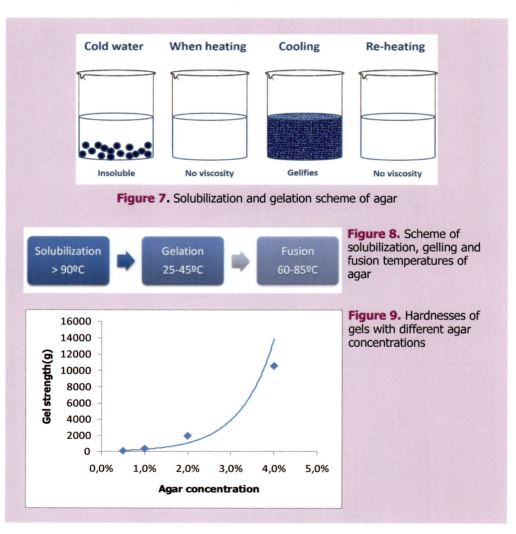

Figure 7. Solubilization and gelation scheme of agar

Figure 8. Scheme of solubilization, gelling and fusion temperatures of agar

Figure 9. Hardnesses of gels with different agar concentrations

Handbook of hydrocolloids

Table 4. Characteristics of gels

	Agar
Gel type	Firm and brittle
Syneresis	Yes, but decreases as it increases the gel strength
Temperature stability	Thermoreversible, high hysteresis
pH stability	Maximum between 5 and 9
Ionic charge of the medium	High concentrations of salt prevent gelation
Presence of alcohol	It increases gel strength
Synergies	Locust bean gum

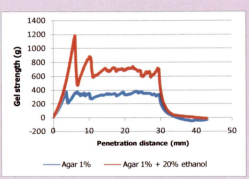

Figure 10. Difference in the hardness of two gels of agar 1%, one with a 20% added ethanol

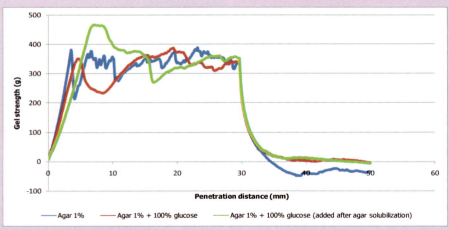

Figure 11. Graphic of agar 1% gel strength, with and without added glucose

Figure 12. Agar gel and detail of the section

Agar

Synergies or incompatibilities

- Agar + locust bean gum: agar has synergies with locust bean gum, an hydrocolloid which acts as a thickener because it cannot make a gel by itself. In Figure 13, you can see the hardness graph of gels formed with different proportions of agar:-garrofin against an agar gel. It is observed that keeping all of them at a concentration of hydrocolloids at 1%, with a ratio 4:1 agar:garrofinit doubles the strength of the gel, as it almost triples in a 1:1 mixture.

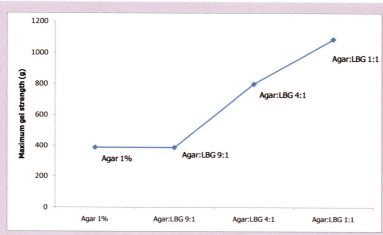

Figure 13. Increase in gel strength for synergy between LBG and agar

Examples of applications in food industry

- Gelling agent in desserts, dairy and confectionery, vegetable alternative to gelatin (derived from pig).

CHEESECAKE WITH STRAWBERRY AND RASPBERRY CREAM

Ingredients: pasteurized cheese (pasteurized milk, sugar, proteins of milk, stabilizer (carboxy methyl cellulose, sorbitol)), gelling agents (**agar**, tara gum), preservative (potassium sorbate)), strawberry and raspberry cream (15%) (strawberry and raspberry puree, sugar, glucose syrup, water, natural flavouring, colour).

ASSORTED FRUIT GUMS

Ingredients: sugar, corn syrup, corn starch, juice of concentrates fruits (2,9%) (lemon (0,81%), apple (0,65%), raspberry (0,62%), blackcurrant (0,34%), cherry (0,19%), pineapple (0,19%), orange (0,1%)), acidity regulator (citric acid, lactic acid, malic acid, calcium citrate, sodium citrate), gelling agent (pectin, **agar**), flavourings and colours.

SHEEP'S MILK JUNKET

Ingredients: sheep milk, stabilizer (**agar**), salt and microbial coagulant.

Handbook of hydrocolloids

- Stabilizer in pastry or fillers for their high hysteresis.

- Stabilizer in sorbets and ice creams to avoid recrystallization.

- Water retention in sterilized meat products

CHOUX PASTRY FILLED WITH COFFEE CUSTARD

Ingredients: water, sugar, eggs, flour, milk powder, butter (4,9%), modified starch, cream (2,5%), glucose syrup, non-hydrogenated vegetable fat (palm, palm kernel), skimmed milk powder, instant coffee (0,9%), stabilizers (carrageenan, glycerol), gelling agents (diphosphates, sodium alginate, **agar**), salt, emulsifier (citric esters of mono- and diglycerides of fatty acids, lecithin), wheat gluten, preservative (potassium sorbate), flour treatment agent (ascorbic acid).

 Vegetable gelatin

Agar is also known as vegetable gelatin, and it is used in many recipes as a substitute for animal gelatin since it has similar characteristics. However, in the food industry, it is more common to use the kappa carrageenan because gelling and melting temperatures are more similar to the animal gelatin.

5 E-407 CARRAGEENAN

E 407 Carrageenan (CAS Numbers: 11114-20-8(kappa), 9062-07-1 (iota), 9064-57-7 (lambda))

E-407a Processed Eucheuma Seaweed or PES (CAS Number: 9000-07-1)

Other denominations: carrageenin

Handbook of hydrocolloids

Origin

It is obtained from the intercellular material of rhodophyceae algae (red algae) (Figure 1).

There are four types of carrageenan, each with their own properties. Those are: kappa I (κ), kappa II (κ), iota (ι) and lambda (λ (Figures 2, 3 , 4 and 5)). Depending on the species of red algae where extraction is made, the obtained products will be rich in a type of carrageenan or the other (Table 1).

Composition and chemical structure

Carrageenan is a heteropolysaccharide made by galactose polymers more or less sulfated. The main molecules are: D-galactose and 3,6-anhydrous-D-galactose (3,6 AG), with a higher or lower sulfation degree. The four types of carrageenan are differentiated by the content of 3,6 AG and ester sulfate in the D-galactose, which influences the solubility of the molecules. Thus, a higher ester sulfate percentage increases the solubility of carrageenan and a bigger proportion of 3,6 AG increases the temperature required for the molecule to solubilize (Table 2).

The E-407a, also known as PES (Figure 6), has a different extraction process than the rest of carrageenan, much simpler and cheaper since the carrageenan does not purify at all. This results in a product with a higher proportion of fibers and insoluble materials which are expressed as AIM (Acid Insoluble Matter) and that can be between 8-15% of the final weight, while the refined carrageenan should not exceed 2% , as it appears in the Regulation No 231/2012 of 9 March 2012 laying down specifications for food additives listed in Annexes II and III to Regulation (EC) No 1333/2008 of the European Parliament and of the Council.

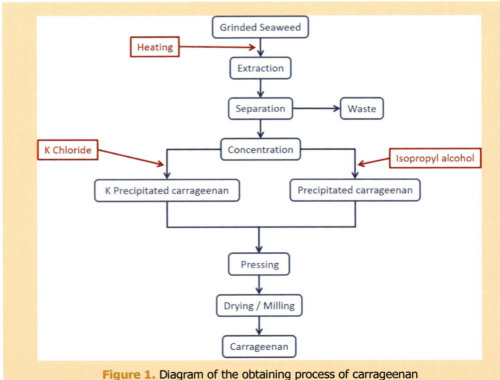

Figure 1. Diagram of the obtaining process of carrageenan

Carrageenan

E-407

Table 1. Extraction of different types of carrageenan according to origin

Species	Chondrus sp.	Euchema sp.		Gigantina sp.		Furcellaria sp.
Subspecies	Ch. crispus	E. cottonii	E. spinosum	G. stellata	G. radula	F. fastigiata
Origin	North Atlantic	Philippines		Spain, France	Argentina, Chile	Denmark, Canada
Carrageenan type	λ, κ	κ	ι	λ, κ		λ, κ

Table 2. Differences in composition between different types of carrageenan

Carrageenan	Kappa I	Kappa II	Iota	Lambda
Ester sulfate	20-25%	25-28%	32%	35%
3,6-anhydrous-D-galactose	35-40%	35-40%	30%	Almost non-existent

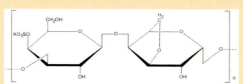

Figure 2. Kappa I carrageenan molecule

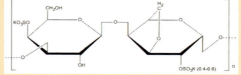

Figure 3. Kappa II carrageenan molecule

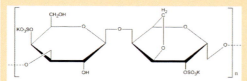

Figure 4. Iota carrageenan molecule

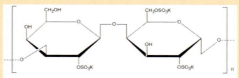

Figure 5. Lambda carrageenan molecule

Figure 6. E-407a gel, Processed Eucheuma Seaweed (PES)

> ⓘ Commercially, it is easier to find mixtures of two or more types of carrageenan. For example, a kappa carrageenan may contain a percentage of iota and lambda, so it may have the characteristics of both carrageenans.

Properties of solutions

Solubility and solutions preparation

All types of carrageenan are soluble when exposed to heat. Lambda carrageenan and some iota are also soluble in cold conditions. They tend to get wet on the surface when they come into contact with liquids and form lumps, so it is highly recommended premixing with other powder products in formulation and applying mechanical work for a correct dispersion (Table 3).

Despite the common traits, each type of carrageenan has differences in solubility, as summarized in table 4.

Factors affecting the properties of solutions

Kappa I and II and iota carrageenan have no effect on the viscosity of the medium once solubilized. However, with lambda carrageenan you can see differences in viscosity, basically related to the concentration of the hydrocolloid and temperature of the medium (Table 5).

Table 4. Main differences in carrageenan solubility

Type of carrageenan	Kappa I and II	Iota	Lambda
Cold conditions	Insoluble	Insoluble, except their sodium salts	Soluble
Presence of ions	Needs K$^+$	Needs Ca^2	Does not affect
High [NaCl]	Decreases solubility	Does not affect	Does not affect
High [sugar]	It is not affected if the product is solubilized first, and the sugar is added afterwards.	Slightly soluble	Does not affect

Table 5. Factors affecting the properties of solutions

Factors	Effects on viscosity	Observations
Molecular weight / polymerization degree	Directly proportional	
Concentration	Directly proportional	It does not affect kappa or iota
Temperature	In lambda, it is inversely proportional. It does not affect kappa and iota.	An increase in temperature causes a decrease in viscosity
pH	It is stable over a wide pH range	
Ionic charge	Does not affect	
Mechanical work	Pseudoplastic behaviour	It decreases the viscosity as it increases the shear force

Carrageenan

E-407

Table 3. Solubility and solutions preparation / +++ (high) ++ (middle) + (low)+ (bajo)

Carrageenan	Solubility in cold water	Solubilization temperature	Premix	Shear force	Air incorporation	Increase viscosity of medium	Precipitates with alcohol
Kappa I	Insoluble	++	++	++	No	+	No
Kappa II	Insoluble	++	++	++	No	++	No
Iota	The most insoluble	++	++	++	No	+	No
Lambda	Soluble	Hot and cold	+++	+++	No	+++	No

Table 6. Characteristics of gels

Type of carrageenan	Kappa I	Kappa II	Iota	Lambda
Gel type	Firm and crisp	Firm and elastic	Weak, elastic and cohesive	It does not gel, gives viscosity in cold and hot
Syneresis	Yes, lower in presence of locust bean gum and bigger with KCl	Moderate, less than kappa I	It does not present	
Temperature stability	Thermorreversible, unstable to freezing and defreezing	Thermorreversible, unstable to freezing and defreezing	Thermorreversible, stable to freezing and defreezing	
Reactivity in milk	Moderate	Very high	High	High
Synergies	KCl, Konjac and LBG	k-caseins of milk	Ca^{2+} of dairy products	It does not present

Functionality

Gelation mechanism

Kappa and iota carrageenan gels are formed during cooling after solubilization in hot. They are thermoreversibles and present some hysteresis, a difference of 10 to 20°C between gelling and fusion temperatures (Figure 7). During cooling, double-helix molecular structures are formed that will link between them in the presence of each carrageenan characteristic ion, to form the three-dimensional structure of the gel.

Although the gelation mechanism is quite similar, each type of carrageenan has differences. For instance, kappa carrageenan depends on K^+ ions in the medium for gelation, and iota needs Ca^{2+} ions (Figure 9).

Lambda carrageenan cannot form a gel. In fact, it only adds viscosity to the medium in both cold and hot environments and is not affected by the presence of cations (Figure 8).

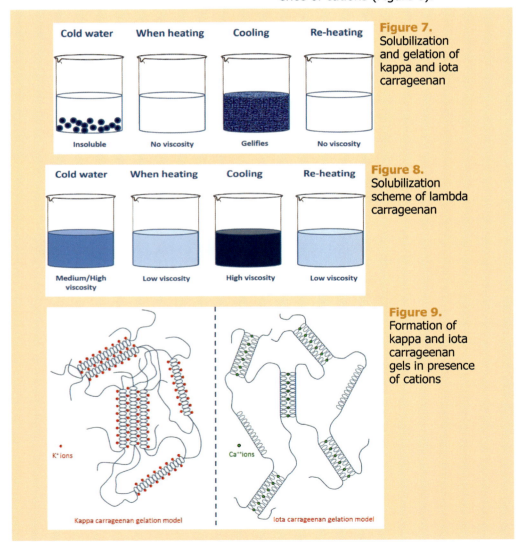

Figure 7. Solubilization and gelation of kappa and iota carrageenan

Figure 8. Solubilization scheme of lambda carrageenan

Figure 9. Formation of kappa and iota carrageenan gels in presence of cations

Carrageenan

E-407

Characteristics of gels

The different carrageenan types make gels that end up having some of the same characteristics. For instance, they are all thermoreversible and present some hysteresis at solubilization, gelling and melting temperatures. However, they can also form gels with different appearances and properties.

The kappa I carrageenan makes firm and brittle gels, with syneresis, that are not stable to freeze-defreeze processes. Depending on the purity degree of the product, they are more or less translucent, although they clarify in the presence of sugar. In the presence of potassium salts, most commonly KCl, the hardness of these gels is increased, but also the syneresis and gelling temperature (Figure 10). Most kappa carrageenans that are currently on the market are of type I and are standardized with a small amount of KCl.

Gels formed by kappa II carrageenan are firm but elastic, without syneresis, thermoreversibles and presenting a great reactivity with the κ-casein of milk. Note that most products of kappa carrageenan in the market are composed of a mixture of two types of kappa, without specifying the proportion of each one.

Processed eucheuma seaweed or PES gels are quite similar to kappa carrageenan. However, due to a larger amount of insoluble matter present, they are completely opaque (Figure 11).

Iota carrageenan gels (Figure 12) are weaker than the kappa. They are cohesive, elastic and can be reformed after broken. They are transparent, stable to freeze and defreeze and do not have syneresis.

Figure 10. On the right is shown an increase of the syneresis in a kappa gel, standardized with KCl compared with a non-standardized one (left)

Figure 11. Difference of transparency in a kappa carrageenan gel (left) and a PES one (right)

Figure 12. Iota carrageenan gels

Handbook of hydrocolloids

Synergies or incompatibilities

- Kappa carrageenan + locust bean gum (LBG) / kappa carrageenan + konjac: the two hydrocolloids have a synergy with kappa carrageenan, giving harder and elastic gels with a significant decrease of syneresis. This hardness increase is much more pronounced with konjac, as shown in Figure 13, where the maximum hardness of a 1% kappa carrageenan gel is multiplied by 5 by replacing 0,3% of kappa with konjac.

- Kappa carrageenan and κ-casein interaction: in dairy mediums where the pH is above the isoelectric point of proteins, there is an interaction between the negative charges of the ester sulfate groups of kappa carrageenan with positive surface charges of the casein micelles, forming an undetectable three-dimensional network that can stabilize the aqueous phase of all kinds of dairy products (Figure 14). Very low doses of kappa carrageenan are needed to form it, and it is favored by K^+ and Ca^{2+} ions present in the dairy medium. It is very useful to facilitate the suspension of cocoa particles in milk drinks.

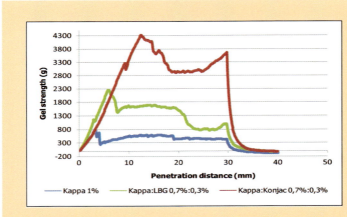

Figure 13. Effects on hardness of kappa gels in synergy with LBG and konjac.

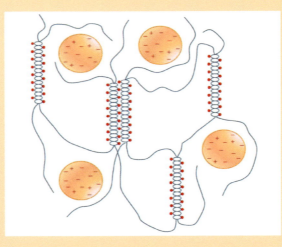

Figure 14. Scheme of kappa carrageenan and κ-casein interaction

Carrageenan

E-407

> ℹ️ Although the interaction of kappa carrageenan with milk caseins is usually described in the literature, this interaction also occurs with proteins of vegetable origin, so the application of the hydrocolloid as a suspensor of particles like cocoa is extended to all types of vegetable drinks.

Examples of applications in food industry

- Water retention in meat or fish, especially in injected products or with a massage.

COOKED HAM

Ingredients: ham (85%), salt, dextrose, sugar, stabilizers (diphosphates and **carrageenan**), preservative (sodium nitrite), antioxidant (sodium ascorbate), spices, flavouring, flavour enhancer (monosodium glutamate).

- Improve the slicing of cooked products.
- Stability and retention of water in cream

UHT WHIPPING CREAM

Ingredients: cream of 35% F.M, stabilizer (**carrageenan**).

- Suspension of cocoa particles in milk or vegetable beverages after formation of three-dimensional network by interaction with proteins.

CHOCOLATE-FLAVORED SOY BEVERAGE

Ingredients: water, peeled soy beans (11%), sugar, fat-free cocoa (1%), acidity regulator (monopotassium phosphate), salt, flavouring, stabilizers (calcium carbonate, **carrageenan**), vitamin B2 (riboflavin) and vitamin B12.

UHT COCOA DRINK

Ingredients: whole milk, skimmed milk, whey, sugar, fat-free cocoa (1,5%), stabilizers (sodium phosphates, **carrageenan**) and flavouring.

- Gelling agent in desserts like pudding or gelatin, alone or in combination with locust bean gum to take advantage of the synergy.

CURD PREPARATION

Ingredients: starch, fructose, stabilizer (**carrageenan**), salt, curdling agent and flavourings.

STRAWBERRY JELLY

Ingredients: water, sugar, dextrose, stabilizers (**carrageenan**, locust bean gum and calcium chloride), acidity regulators (citric acid, sodium citrate and potassium citrate), flavourings and colour (carmines).

Handbook of hydrocolloids

> Using carrageenans as additives in the food industry has an advantage at the labeling, unlike other hydrocolloids, since it is possible to use in the same product various types of carrageenan (iota, kappa or lambda) and declare them once under the name "carrageenan" or E-407.

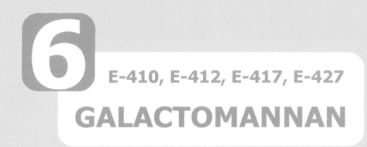

6 E-410, E-412, E-417, E-427
GALACTOMANNAN

E-410 Locust bean gum (CAS Number: 9000-40-2)

Other denominations: LBG, Farine de caroube

E-412 Guar gum (CAS Number: 9000-30-0)

E-417 Tara gum (CAS Number: 39300-88-4)

E-427 Cassia gum (CAS Number: 51434-18-5)

Handbook of hydrocolloids

General introduction

Galactomannan is the generic name used to encompass those polysaccharides of high molecular weight; they are each made up of a chain of mannose united by β (1-4) bonds, which have ramifications of galactose molecules linked by α (1-6) bonds (figure 1).

Galactomannans are non-onic heteropolysaccharides, related to water.

Within the group of galactomannans there are four different types of additives, whose molecules are represented on table 1.

Features in common

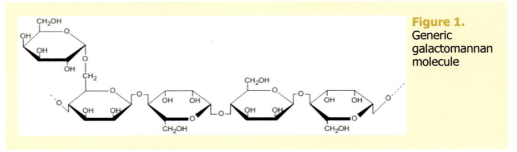

Figure 1. Generic galactomannan molecule

Table 1. Molecules of different galactomannans

Additive	Molecule
E-410 Locust bean gum	
E-412 Guar gum	
E-417 Tara gum	
E-427 Cassia gum	

Galactomannan

- They are obtained from the endosperm of vegetable seeds. Although they come from different backgrounds, all these galactomannans are obtained from the seeds of the fruits of some trees, shrubs and plants.

- They have the same spatial structure; they are composed of a linear chain of mannose with ramifications of galactose.

- They function as thickeners. Although they differ in terms of the use and application, all of these molecules function as thickeners, providing viscosity to the aqueous medium they are added to. They have a pseudoplastic behaviour with values of viscosity between 2000-5000 cps. This value will vary depending on climatic conditions the crops had received, the type of harvesting, the manufacturer and the conditions of use/application in formulation.

- They are partially hydrolyzed in acidic medium.

Differential characteristics

- Ratio mannose:galactose. Each galactomannans has a different proportion between the numbers of molecules of mannose on the branching galactose. Therefore, locust bean gum has a ratio 4:1, guar gum 2:1, tara gum 3:1 and gum cassia 5:1.

- Soluble in hot or cold conditions. The fewer branches this molecule has, the easier and quicker its solubility in ambient temperature will be. Therefore they are considered to be more soluble in cold conditions, providing viscosity without the need of heat.

- Each one is obtained from different vegetable sources (table 2)

- Different organoleptic characteristics. Although they are all thickeners, and have a pseudoplastic behaviour, you can see certain organoleptic and palatability differences depending on viscosity provided by dose and application.

After this introduction to galactomannans, the peculiarities of each of them will be explained, commenting on the origin, composition and chemical structure. We will also talk about them as a whole when explaining the properties of solutions, the form of preparation, synergy, functionality and application examples. That way, it is easy to understand them as a functional set, while we know their different characteristics by origin, needs of use and work.

Table 2. Different origins of galactomannan

Additive	Name	Origin	Scientific name
E-410	Locust bean gum	Carob tree	*Ceratonia siliqua*
E-412	Guar gum	Guar gum plant	*Cyamopsis tetragonoloba*
E-417	Tara gum	Tara plant	*Caesalpinia spinosa*
E-427	Cassia gum	Cassia plant	*Cassia tora, Cassia obtusifoli*

E-410 LOCUST BEAN GUM

Origin

It is a galactomannan obtained from the endosperm of the seed of the fruit of the carob tree, a tree of the Fabaceae family (figure 2).

Composition and chemical structure

Ratio between mannose and galactose of locust bean gum is 4:1 (figure 3). That means that for every 4 molecules of mannose, there is a single branch of galactose. It is one of the less branched galactomannans.

This description is only indicative and gives an idea of the order of branching of the molecules obtained from the seed of the carob tree. It doesn't mean that the molecular structure is as strictly ranked, but in general, it is set at that ratio.

It is necessary to heat a dispersion of locust bean gum to achieve the correct solubility and make it functional as a thickener.

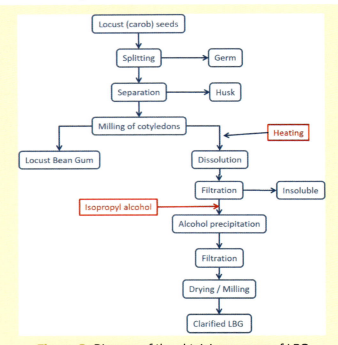

Figure 2. Diagram of the obtaining process of LBG

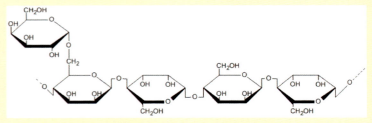

Figure 3. LBG molecule

Galactomannan

E-412 GUAR GUM

Origin
It is obtained from the endosperm of the fruit of an annual plant of the legume family, known as guar plant (figure 4).

Composition and chemical structure
The ratio between mannose and galactose of the guar gum is 2:1 (figure 5). It is the molecule that has more ramifications within the group of galactomannans. Therefore, it has the capacity to more easily capture the water in the medium and hydrate itself, which facilitates its functionality and provides viscosity in an already cold medium.

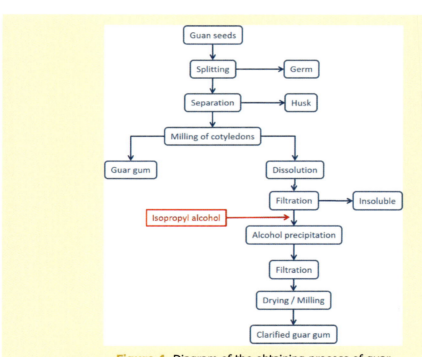

Figure 4. Diagram of the obtaining process of guar gum

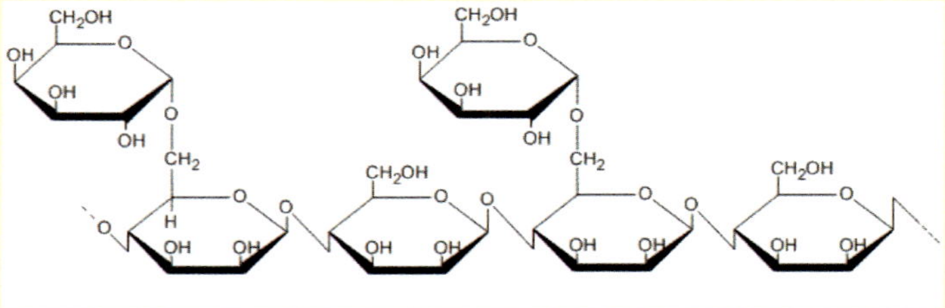

Figure 5. Guar gum molecule

Handbook of hydrocolloids

E-417 TARA GUM

Origin

It is extracted from the endosperm of the fruit seed of a tree in the Fabaceae family called tara (figure 6).

Composition and chemical structure

In the case of the tara gum, the mannose and galactose molecular ratio is 3:1 (figure 7). Meaning that for every 3 linear molecules of mannose, a branch of galactose is located in the main chain. It is an even mix between locust bean gum and guar gum, i.e., it is able to hydrate and give viscosity in cold temperatures, but its maximum solubility and functionality is achieved after a period of heating between 75-95ºC.

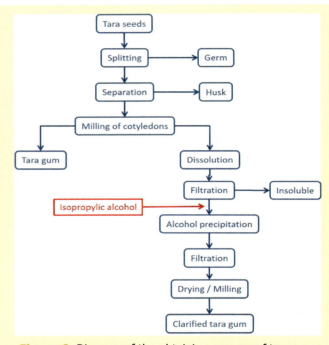

Figure 6. Diagram of the obtaining process of tara gum

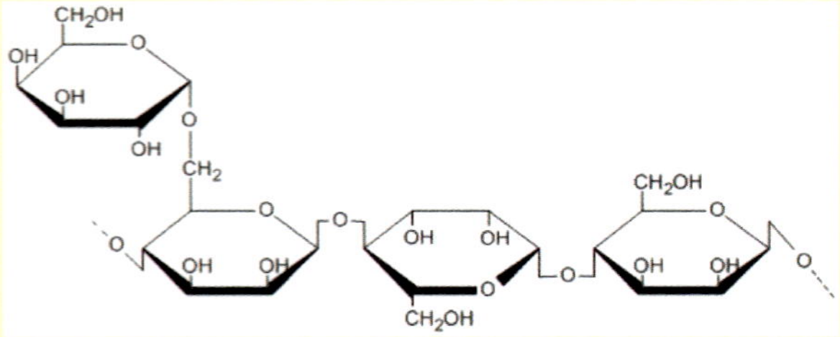

Figure 7. Tara gum molecule

Galactomannan

E-427 CASSIA GUM

Origin

It is obtained from the endosperm of the seed found in an annual plant, known as cassia, of the Fabaceae family (figure 8).

Composition and chemical structure

It is a high molecular weight polysaccharide, and the mannose and galactose ratio is 5:1 (figure 9). It is, therefore, the least branched galactomannan of the whole group, even less than locust bean gum.

Gum (E-427) cassia is rarely used as a thickener in formulation, since it is easier to find the locust bean gum and the application results are similar.

The cassia gum is insoluble in cold conditions, requires heating to solubilize the molecule and functions as a thickener.

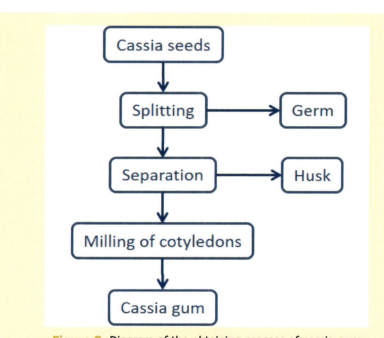

Figure 8. Diagram of the obtaining process of cassia gum

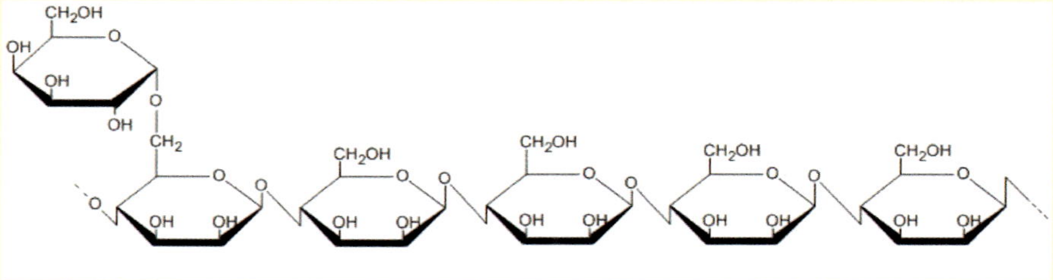

Figure 9. Cassia gum molecule

Handbook of hydrocolloids

Properties of solutions

Solubility and solutions preparation (table 3)

Galactomannans are classified as two groups according to their solubilization temperature.

- **Galactomannan functional in hot conditions:** those that to provide viscosity need a heating stage. Both the LGB (E-410) and cassia gum (E-427) are insoluble in cold or at room temperature. It is necessary to reach a high temperature of about 80-90ºC to achieve the total hydration and dissolution of the product and, therefore, efficiency and functionality providing viscosity. During the warm-up you can see a slight increase in viscosity (depends on the used concentration), but it won't be until cooling the solution that we will be able to see the real viscosity change in the product.

- **Galactomannan functional in cold conditions:** are those that do not need to be heated to provide viscosity to the environment. Guar gum (E-412) and tara gum (E-417) are both found in this group. It is possible to improve their performance by passing through a heating stage or a long-time hydration in cold conditions.

Factors affecting the properties of solutions (table 4)

Once hydrated properly, its viscosity can be affected by several factors:

- **Molecular weight and polymerization degree:** the longer the string of galactomannan, the higher viscosity will give to the medium.

- **Temperature:** at high temperatures the viscosity is lost, however it can be recuperated by cooling the product down.

- **pH:** these molecules are sensitive to acidic pH, since it hydrolyzes them partially and decreases the initial viscosity.

- **Mechanical work:** they are thickeners with pseudoplastic behavior, i.e., agitation reduces the viscosity by rearranging the molecules. However, the viscosity is regained after rest.

Table 4. Factors affecting the properties of solutions

Factors	Effects on viscosity	Observations
Molecular weight/ polymerization degree	Directly proportional	
Concentration	Directly proportional	
Temperature	Inversely proportional	An increase in temperature causes a decrease in viscosity
pH	Acid pH hydrolyzes the molecule	Viscosity decreases in an acidic environment
Ionic charge	Do not affect	
Mechanical work	Pseudoplastic behavior	It decreases the viscosity as it increases the shear force

Galactomannan

E-410
E-412
E-417
E-427

Table 3. Solubility and solutions preparation/ + (high) ++ (middle) + (low)(*it improves solubility with the passage of time)

Galactomannan	Solubility in cold water	Solubilization temperature	Premix	Shear Force	Air incorporation	Increase viscosity of medium	Precipitates with alcohol
Locust bean gum	No	+++	++	+	No	No	Yes
Guar gum	+++	+	+++	+++	+++	++	Yes
Tara gum	++	+(*)	+++	++	++	+	Yes
Cassia gum	No	+++	++	+	No	No	Yes

Handbook of hydrocolloids

Functionality

Galactomannans are thickeners; therefore, they only provide viscosity to the medium. They do not make a gel on their own, but they can produce a gelation in combination with other hydrocolloids. They may be sensitive to freeze-defreeze cycles.

Figures 10, 11 and 12 show the solubilization schemes of different galactomannans to achieve their functionality.

Synergies or incompatibilities

- Galactomannan+ xanthan gum: the synergy between galactomannan and xanthan gum causes a significant increase in viscosity, causing it, in some cases, to form cohesive, elastic soft gels and without syneresis. Mixtures prepared in cold temperatures give more rugged textures, with texture behaviour among gel and highly viscous solution (figure 13). Mixtures prepared in hot temperatures result in smooth, bright and more structured gels (figure 14).

- LBG + kappa carrageenan: LBG manages to eliminate the syneresis of the kappa carrageenan by at the same time achieving an increase in hardness of the gel, certain elasticity and greater resistance (figure 15).

- Galactomannan + starches: is described as a synergistic action between galactomannan with the fraction of amylose in starch, which minimizes retro gradation and delays the hardening of pastries and bakery products.

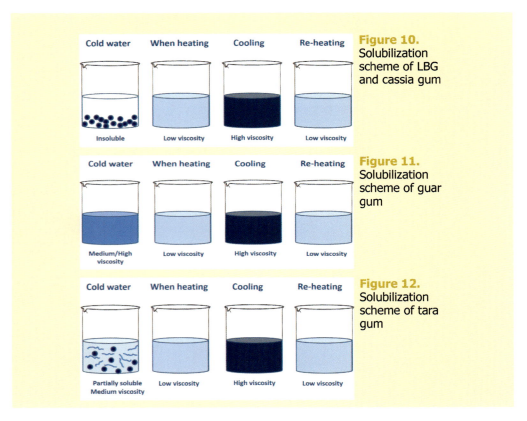

Figure 10. Solubilization scheme of LBG and cassia gum

Figure 11. Solubilization scheme of guar gum

Figure 12. Solubilization scheme of tara gum

Galactomannan

E-410
E-412
E-417
E-427

Figure 13. Synergies between galactomannan and xanthan gum. Preparation of the mixture in cold conditions

Figure 14. Synergies between galactomannan and xanthan gum. Preparation of the mixture in hot conditions

Figure 15. Syneresis reduction in a kappa carrageenan gel with LBG (right) compared to a kappa carrageenan gel (left)

Handbook of hydrocolloids

> When talking about the synergies of galactomannans, most often it is found described for LBG and guar gum, because they are the most frequently used. Even though, as they all have very similar molecular structures, they match in the synergistic behaviour with more or less marked results according to the combination and mode of application.

Examples of applications in the food industry

- Thickener in sauces, creams and condiments.

TUNA WITH TOMATO

Ingredients: tuna (68%), tomatoes (21%), sunflower oil, sugar, wine vinegar, salt, stabilizer (**guar gum**)

BARBECUE RIBS

Ingredients: pork ribs (86%), water, dextrose, tomato, salt, sugar, stabilizer (modified corn starch), extract of yeast, emulsifier (polyphosphate), acidity regulator (citric acid and sodium acetate), spices, smoke flavouring (flavoring substances, maltodextrin and stabilizer (arabic gum)) and other flavourings (soybean). Barbecue sauce (14%) (water, vinegar, sugar, stabilizer (modified corn starch and **guar gum**), colour (caramel ammonium)

- Contribution of creaminess and smoothness by increasing the viscosity in stuffings, cocoa cream, pastry cream, etc.

MEAT CANNELLONI

Ingredients: pork 17%, wheat semolina, tomato puree, wheat flour, vegetables (onion, carrot), melted cheese (contains milk) (cheese, butter, milk proteins, potato starch, emulsifying salts (polyphosphate, potassium phosphate, sodium citrate), salt, stabilizer (**locust bean gum**), colour (annato)), semi-skimmed milk powder, margarine (oils and fats from palm and sunflower, water, emulsifier (mono and diglycerides of fatty acid), salt, acidity regulator (citric acid), preservative (potassium sorbate), flavourings, colour (beta-carotene), breadcrumbs (wheat flour, water, salt, yeast), sunflower oil, modified starch, salt, white wine, spices (black pepper, nutmeg).

- Delays the hardening for retrogradation of starch in baked goods and pastries

- Thickener and contributor in texture of spread products, cheese spreads, etc.

Galactomannan

E-410
E-412
E-417
E-427

PASTEURIZED WHITE CHEESE WITH PINEAPPLE

Ingredients: milk, cream, blend of papaya/almonds for decoration 15% (50% almonds, 50% candied papaya (papaya 75%, sugar cane 17,5%, rice flour 7,4%, acidity regulator (citric acid)) pineapple 7,5%, pineapple concentrate 4,6%, sugar 2.5%, milk protein, gelatin, salt, citrus fiber, acidity regulator (lactic acid), stabilizer (**locust bean gum**), preservatives (sorbic acid)

- Gelling agent in desserts in combination with kappa carrageenan for the synergy with LBG.

STRAWBERRY CREAMY DESSERT

Ingredients: modified starches, sugar, powdered milk, milk solids, sweetener (cyclamate), antioxidant (citric acid), stabilizer (**locust bean gum**, carrageenan), flavouring and colour (cochineal red A, carmoisine)

- Stabilizer in emulsified products. Helps the stability of the emulsion to provide viscosity in the continuous phase.
- Improves texture in yogurt smoothies to provide creaminess without using cream or minimizing their use.
- Improves texture and slicing of cooked meat products (like ham), applied along with kappa carrageenan, to achieve a better synergy and less syneresis.
- Contributes in the viscosity of sauce, among the pieces of meat in pet food products
- Stabilizers in ice cream.

STRAWBERRY FLAVOUR ICE CREAM

Ingredients: reconstituted skimmed milk powder, water, sugar, hydrogenated vegetable fat, milk solids, strawberry pulp (7,1%), glucose syrup, emulsifier (mono and diglycerides of fatty acids), stabilizers (**guar gum, locust bean gum,** carrageenan), flavouring, acidity regulator (citric acid), colour (cochineal)

SORBETS WITH LEMON, ORANGE, APPLE, STRAWBERRY, TANGERINE AND LIME-LEMON

Ingredients: water, pulp and based concentrates fruit juice (18,5%) (lemon (35,7%), orange (18,4%), apple (17,8%), strawberries (13,5%), tangerine (13,5%), lime (1,1%)), sugar, glucose and fructose syrup, stabilizers (**guar gum**, carrageenan, **locust bean gum**), emulsifiers (mono and diglycerides of fatty acids), acidity regulator (citric acid), flavourings, colour (riboflavin, β-carotene, carminic acid, curcumin and copper complexes of chlorophyllins).

- Thickener in drinks to increase mouthfeel.

NON-SPARKLING REFRESHING FLAVORED DRINK

Ingredients: water, glucose-fructose syrup, sugar, fruit juice from concentrate (5%) (orange, tangerine, grapefruit and lime), acidity regulator (citric acid), preservatives (polyphosphate, potassium sorbate), stabilizer (**guar gum**), vitamins, emulsifier (sodium octenilsuccinate of starch), natural flavourings.

7 E-413 TRAGACANTH GUM

E-413 Tragacanth gum (CAS Number: 9000-65-1)

Other denominations: adragante gum, alquitira gum, dragon gum

Origin

Tragacanth gum is obtained from the ooze of a perennial shrub of the Middle East, *Astragalus gummifer* and also, but not as often, from *Astragalus microcephallus* (fam. *Leguminosae*) (figure 1).

The word "tragacanth" comes from the Greek "tragos" (goat) and "akantha" (horn), and refers to the form of ribbon twisted spiral that adopts the rubber when it exudes from the wounds of the plant.

Composition and chemical structure

Tragacanth gum is an anionic branched heteropolysaccharide formed by two fractions: basorine and tragacanthine.

- **Basorine:** it constitutes the 60-70% of the total gum. Formed by units of galacturonic acid, xylose, galactose, arabinose and small amounts of rhamnose. It may also contain small amounts of starch, cellulose and protein. It also contains some methyl groups (similar to pectin). It hydrates in cold water but does not manage to solubilize unless bringing the solution to boil (figure 2).

- **Tragacanthine:** formed by a non-ionic and highly branched arabinogalactane composed by units of galactose, xylose and arabinose. It easily solubilizes in cold water (figure 3).

The composition varies quite according to crops, origins and species. An increase of galacturonic acid and methyl groups imparts higher viscosity to solutions, while an increase of galactose and arabinose decreases it.

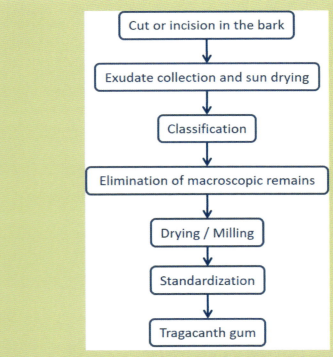

Figure 1. Diagram of the obtaining process of tragacanth gum

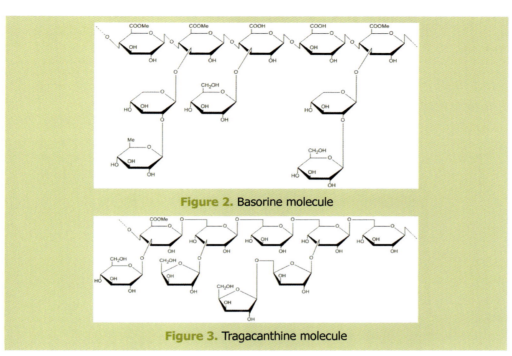

Figure 2. Basorine molecule

Figure 3. Tragacanthine molecule

Properties of solutions

Solubility and solutions preparation

Preparing tragacanth gum solutions does not usually present difficulties if they are prepared in cold water, since they don´t tend to form lumps difficult to disperse or to incorporate lots of air when stirring (Table 1).

Solutions of tragacanth gum are pseudoplastic and slightly acidic (pH 5-6). Tragacanth in cold water is simply hydrated. It is necessary to increase the temperature to a boiling point in order to completely solubilize the solution.

Tragacanth gum is insoluble in alcohol and organic solvents. If the solution has been previously prepared in an aqueous solution, it can tolerate low concentrations of alcohol.

Factors affecting the properties of solutions (Table 2)

- **Temperature:** tragacanth gum is insoluble in cold water and, therefore, it is necessary to heat solutions up to 80°C to get the correct solubility. Once properly solubilized, the viscosity of the solutions is inversely proportional to the temperature this is exposed to (figure 4).

- **Ionic strength:** being an anionic molecule, a high concentration of ions in the medium can influence its correct hydration and solubilization.

- **pH:** tragacanth gum is stable in acid media.

Table 1. Solubility and solutions preparation / +++ (high) ++ (middle) + (low)

Solubility in cold water	Solubilization temperature	Premix	Shear force	Air incorporation	Increase viscosity of medium	Precipitates with alcohol
+	+++	++	++	No	+	Yes

Handbook of hydrocolloids

Table 2. Factors affecting the properties of solutions

Factors	Effects on viscosity	Observations
Molecular weight / polymerization degree	Directly proportional	
Concentration	Directly proportional	
Temperature	Once solubilized, inversely proportional	An increase in temperature causes a decrease in viscosity
pH	Very stable at acid pH. At slightly alkaline pH it partially depolymerizes over time.	The solutions are, in themselves, slightly acidic
Ionic charge	Slightly decreases viscosity	
Mechanical work	Pseudoplastic behaviour	It decreases the viscosity as it increases the shear force

Functionality

Tragacanth gum is a thickener that requires heating to exert its effect. Equal concentration provides less viscosity than other thickeners and is very stable in an acid medium.

Tragacanth gum has the capacity to reduce the surface tension of the solutions, which, together with its thickening character, make tragacanth gum an interesting emulsifier. It's emulsifying ability is similar to that of an emulsifier with an HLB of 11. The emulsifying capacity of tragacanth is due to the presence of protein in its molecule.

Synergies or incompatibilities

- Tragacanth gum + arabic gum: tragacanth gum introduces a slight incompatibility in combination with arabic gum since the viscosity of the mixture is lower than would be expected if used separately.

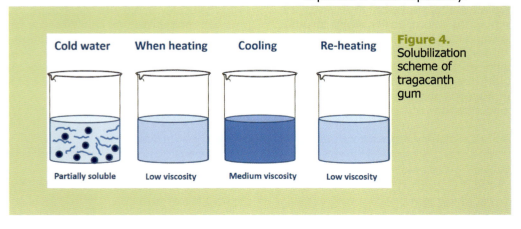

Figure 4. Solubilization scheme of tragacanth gum

Tragacanth gum

E-413

Examples of applications in food industry

- Stable thickener in acidic media

SALAD DRESSING

Ingredients: cream, milk, modified starch, gelatin, acidity regulators (lactic acid, citric acid), onion powder, salt, stabilizers (guar gum, **tragacanth gum**, monoglycerides of fatty acids), spices and preservative (potassium sorbate)

LUMPFISH ROE

Ingredients: lumpfish roe (84%), salt, water, sugar, stabilizer (**tragacanth gum**), acidity regulator (citric acid), preservatives (potassium sorbate and sodium benzoate) and colour (caramel and brilliant black)

- Formation of adherent films that allow hook seeds on the surface of bread and buns,

- Emulsions by lowering the surface tension and giving viscosity stabilizer.

8 E-414 GUM ARABIC

E-414 Gum arabic (CAS Number: 9000-01-5)

Other denominations: Acacia gum, kordofan (name of the Sudan area where it is mainly harvested)

Origin

Gum arabic is obtained from the gummy extract of different types of acacias, especially from *Acacia seyal* and *Acacia senegal* (Fam. *Fabaceae*) (figure 1).

The name "arabic" is due to the fact that the gum was exported from Arabia and distributed all over the world.

Acacia gum drops are processed in two ways: wet and dry. The dry method is simply a physical process of crushing and sieving, which removes most impurities. However, the wet method begins with the dissolution of the gum followed by the elimination of impurities by filtration and/or centrifugation and a final pasteurization, resulting in a product of higher quality and easier solubilization.

Composition and chemical structure

Gum arabic is an anionic branched polysaccharide that incorporates proteins in its molecule (figure 2).

The polysaccharide fraction known as "arabinogalactan" is the most abundant and is composed of molecules of galactose, arabinose, rhamnose and glucuronic acid. Some of the molecules are methylated (Table 1).

The protein fraction consists of serine and hydroxyproline-rich glycoproteins. The amount of glycoproteins in *Acacia senegal* is greater than in *Acacia seyal* (Table 2).

A good way to explain the structure of the gum arabic is the one proposed by Wattle Blossom (Figure 3). He represents the molecule as a support consisting of a long chain of protein that is used as a connection between the units of arabinogalactan.

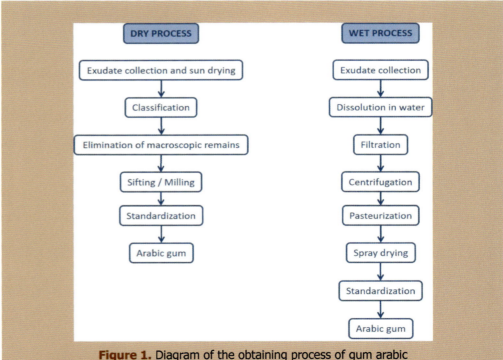

Figure 1. Diagram of the obtaining process of gum arabic

Table 1. Gum arabic average composition

Arabinogalactan (87-98%)	Glycoproteins (2-13%)
Galactose (35-45%)	
Arabinose (25-45%)	
Rhamnose (4-13%)	*Acacia senegal* > *Acacia seyal*
Glucuronic acid (6-15%)	
Some methyl groups	

Table 2. Differences of gum arabic by origin

Acacia seyal - Talha	*Acacia senegal* - Hashab
Drops more brittle, broken	Drops less brittle, rounded
Darker color	Lighter color
More methylated	Less methylated
Arabinose > Galactose	Galactose > Arabinose
Less % protein	More % protein
Dextrorotatory solutions	Levorotatory solutions
Pm ≈ 800.000 D.	Pm ≈ 400.000 D

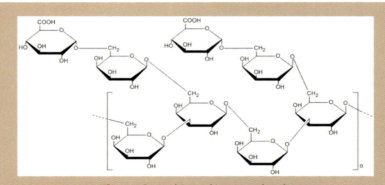

Figure 2. Arabinogalactan molecule

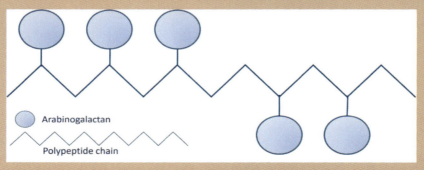

Figure 3. "Wattle Blossom" model for gum arabic (Fencher, 1983)

Properties of solutions

Solubility and solutions preparation (Table 3)

Gum arabic does not provide viscosity when used in usual doses. In equal concentration, solutions of gum arabic are about 600 times less viscous than guar, locust bean gum and xanthan gum (5 mPa.s vs 3000 mPa.s).

This makes it possible to prepare a highly concentrated fluid solution of gum arabic, even 50% (w/w), something that would not be possible with other hydrocolloids. In cases like these, it is best to prepare rather hot solutions, since the air incorporated when stirring is automatically removed.

Factors affecting the properties of solutions (Table 4)

Since gum arabic provides low-viscosity solutions, it is easy to disperse and moisturize in low concentrations. On the other hand, if very concentrated solutions are prepared it can be difficult to get the proper hydration of the product and is often necessary to heat the solution, thus decreasing the viscosity and facilitating the defoaming.

Functionality

The basic characteristics that differentiate the gum arabic from the rest of hydrocolloids can be summarized in three basic properties:

Table 3. Solubility and solutions preparation / +++ (high) ++ (middle) + (low)

Solubility in cold water	Solubilization temperature	Premix	Shear force	Air incorporation	Increase viscosity of medium	Precipitates with alcohol
+++	No	+++	+++	+++	No	No

Table 4. Factors affecting the properties of solutions

Factors	Effects on viscosity	Observations
Molecular weight / polymerization degree	Unimportant except in very concentrated solutions	
Concentration	Directly proportional	Only seen at high concentrations
Temperature	Inversely proportional	An increase in temperature causes a decrease in viscosity
pH	It is stable over a wide pH range	Arabian solutions is, of itself, slightly acidic
Ionic charge	Does not affect	
Mechanical work	Newtonian behaviour	Viscosity measurement does not depend on the shear strength

Gum Arabic

E-414

- **Thickening:** solutions of gum arabic only provide viscosity at a very high concentration ratio. This allows the preparation of syrups with low humidity and low water activity (figure 4).

- **Emulsifying capacity:** gum arabic has an interesting emulsifying capacity due to the presence of hydrophilic polysaccharides (arabinogalactan) and proteins with a high percentage of hydrophobic amino acids (serine and hydroxyproline) (figure 5).

- **Film-forming capacity:** film-forming properties are due to the arabinogalactan fraction. The low viscosity and the possibility of making solutions with high dry extract (up to 50% of gum arabic) facilitate the formation of films once the water evaporates.

Synergies or incompatibilities
- Gum arabic + emulsifiers: numerous emulsifiers can be used in combination with gum arabic to further enhance its abilities.

Examples of applications in food industry
- Emulsifier in fat-soluble products like oleoresins, beta carotene and vitamins (A, D, E, K).

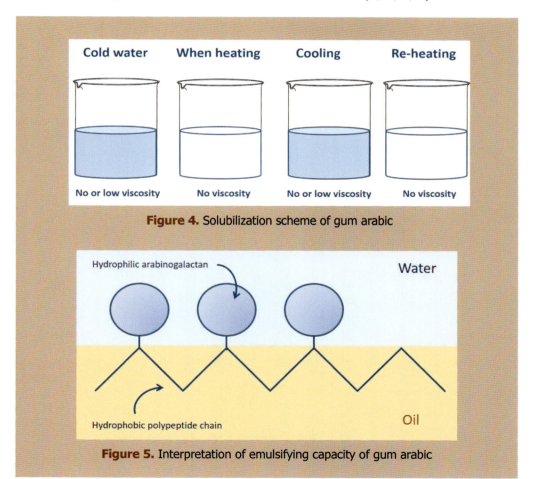

Figure 4. Solubilization scheme of gum arabic

Figure 5. Interpretation of emulsifying capacity of gum arabic

Handbook of hydrocolloids

- Emulsifier in oils used as flavoring, especially in acidic soft drinks.

> **ORANGE SODA**
>
> Ingredients: water, orange juice based on concentrate (5%), added carbon dioxide, acidity regulator (citric acid), natural flavouring, sweeteners (acesulfame K and sucralose), vegetable oil, stabilizers (**gum arabic** and SAIB), colour (beta carotene)

- Beer foam stabilizer.

> **BEER**
>
> Ingredients: water, barley malt, hops, stabilizer (**gum arabic**)

- Adherent of flavorings or salt on snacks and nuts.

> **ROASTED AND SALTED SUNFLOWER SEEDS**
>
> Ingredients: Sunflower seeds, salt, rice flour, stabilizer (**gum arabic**)

- Film-forming agent that forms an anti-humidity and protective barrier against oxidation in snacks and nuts.

> **JAPANESE PEANUTS**
>
> Ingredients: peanuts, wheat flour, modified starch, sugar, soy sauce (hydrolyzed soybean vegetable protein, colour (caramel)), flavor enhancer (monosodium glutamate), iodized salt, stabilizer (**gum arabic**), chili powder

- As a texture agent when used at high concentration in the production of gummies or licorice tablets.

> **SUGAR-FREE LICORICE**
>
> Ingredients: sweeteners (maltitol syrup, steviol glycosides), thickener (**gum arabic**), water, licorice extract (2%), stabilizer (glycerin), natural flavoring, glazing agent (beeswax)

 Envelopes and stamps

As a curious application of gum arabic, it has been used as the glue on stamps and envelopes, when they were not adhesive and should get wet to work. It needed to be a food product (since people hydrate it by licking it), that will easily form film, that could rehydrate itself quickly and with good adhesive properties. Gum arabic met all the requirements.

9 E-415 XANTHAN GUM

E-415 Xanthan gum (CAS Number: 11138-66-2)

Other denominations: santana, xantano

Handbook of hydrocolloids

Origin

Xanthan gum is an exomucopolisaccharide that is obtained by bacterial fermentation of *Xanthomonas campestris* (figure 1). It is used as a thickener in the food industry.

Composition and chemical structure

It presents a complex molecular structure, with a main skeleton identical to the composition of cellulose, composed of D-glucopiranosas united through bonds β (1-4). The main chain has randomly linked ramifications of a trisaccharide composed of two units of D-mannopyranose and a molecule of glucuronic acid. D-mannopyranose units also carry acetyl groups (figure 2).

Xanthan gum adopts a spatial structure that brings sufficient rigidity to be resistant to some medium factors like high temperatures, acidic pH, etc., without being hydrolyzed as it happens to other molecules. However, its high branching facilitates hydration in cold.

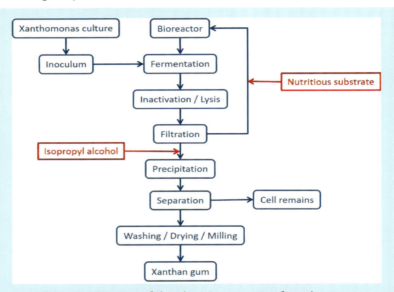

Figure 1. Diagram of the obtaining process of xanthan gum

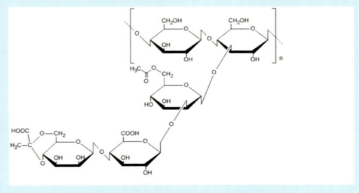

Figure 2. Xanthan gum molecule

Properties of solutions

Solubility and solutions preparation (table 1)

Xanthan gum is hydrated and solubilizes in cold conditions. It is characterized as a hydrocolloid, which gives high viscosity at low concentrations when compared with other thickener additives (figure 3). Moreover, it is one of the most stable hydrocolloids, since it can resist, without hydrolyzing itself, in environmental conditions that other thickeners would not.

Factors affecting the properties of the solutions (table 2)

- **Temperature:** it remains stable at sterilization processes almost without registering changes of viscosity (during the heating process there is low viscosity, but once returned to cool, it recovers its initially).

- **pH:** xanthan gum is very stable to variations in pH of the medium, and can withstand extreme pH between 2.5 and 10. It offers good behavior if it is subjected to acid pH and heat treatment at the same time.

- **Ionic charge:** xanthan gum shows enough stability in media with a high ionic charge, like brines (figure 4).

- **Enzymes:** the presence of enzymes of some foods is another trigger for the loss of viscosity due to hydrolysis by enzymatic degradation. Xanthan gum is resistant to the attack of enzymes like protease, pectinase and cellulase, hemicellulase and amylase.

- **Alcohol:** in media with the presence of alcohol, the behavior of xanthan gum may vary depending on the alcoholic content. Media containing less than or equal to 50% of ethanol allows xanthan gum to remain stable, acting as a thickener, and may even increase its thickening effect (figure 5). However, at concentrations greater than 50-60% of ethanol, xanthan gum is insoluble.

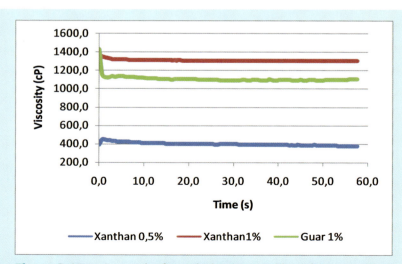

Figure 3. Viscosity graph of two thickeners in cold conditions: xanthan gum (0,5% to 1%) and guar gum (1%) (Rotational viscometer (50 rpm, 60 s, spindle R6))

Handbook of hydrocolloids

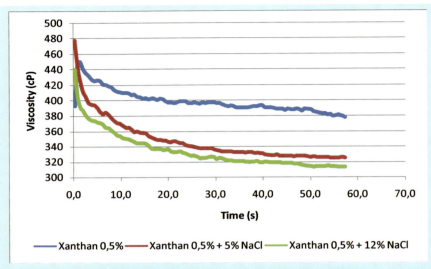

Figure 4. Viscosity graph of xanthan gum at 0,5%, 0,5% in the presence of 5% NaCl and 12% NaCl (rotational viscometer, 50 rpm, 60 s, spindle R6)

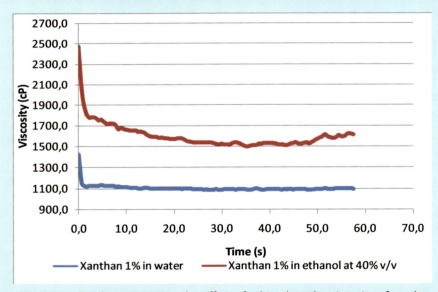

Figure 5. Graph representing the effect of ethanol on the viscosity of a solution of xanthan gum 1% (rotational viscometer (50 rpm, 60 s, spindle R6))

Table 1. Solubility and solutions preparation / +++ (high) ++ (middle) + (low)

Solubility in cold water	Solubilizing temperature	Premix	Shear force	Air incorporation	Increase viscosity of medium	Precipitates with alcohol
+++	cold	+++	+++	+++	+++	Yes, >50%

Xanthan Gum

Table 2. Factors affecting the properties of solutions

Factors	Effects on viscosity	Observaciones
Molecular weight/ polymerization degree	Directly proportional	
Concentration	Directly proportional	
Temperature	Inversely proportional	An increase in temperature causes a decrease in viscosity. Stable at sterilization
pH	It is stable over a wide pH range	
Ionic charge	Low decrease of viscosity, but it is vety stable	
Mechanical work	Pseudoplastic and rheopectic behaviour	It increases the viscosity as it increases the strength of shear prolonged in time
Presence of alcohol	Increases the viscosity in concentrations <50%	Precipitates at concentrations >50%
Enzymes	Stable to most enzymatic attacks (proteases, amylases, cellulases, etc...)	

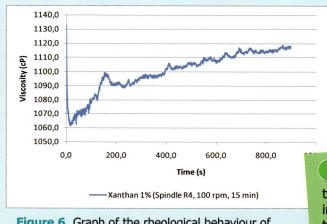

Figure 6. Graph of the rheological behaviour of xanthan gum

ⓘ In some cases, xanthan gum can be described in the literature as thixotropic (decrease in viscosity over time with constant shear force). However, in Figure 6 it shows a rheopectic behaviour.

Handbook of hydrocolloids

Functionality

Xanthan gum works as a thickener in cold conditions, and does not need to be heated to provide viscosity to the medium. That viscosity achieved is very stable (figures 7, 8).

As a thickener, it shows a pseudo-plastic behavior, with rheopectic features. I.e., if a shear force is applied to a solution with a 0,5-1% concentration of xanthan gum, the viscosity decreases (pseudoplastic behaviour), but if such agitation is maintained constantly, during a time, the viscosity increases. Once it ceases to apply the agitation, it takes a while to recover its initial viscosity (rheopectic behaviour).

Synergies or incompatibilities

- Xanthan gum + Galactomannans: as a result of the combination of xanthan gum with each of galactomannans, it gets a synergy that in some cases produces a gel. Its texture depends on the concentration of added hydrocolloid and the ratio between the two gums. It is also very important if the application is made in hot or cold conditions (table 3) (figures 9, 10).

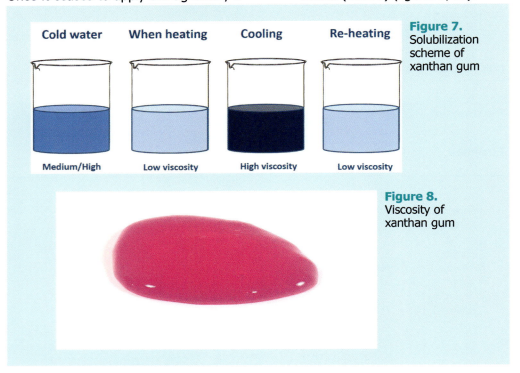

Figure 7. Solubilization scheme of xanthan gum

Figure 8. Viscosity of xanthan gum

Table 3. Synergies between xanthan gum and galactomannan

	Locust bean gum		Guar gum		Tara gum	
	Cold	Hot	Cold	Hot	Cold	Hot
Xanthan gum	Increase of viscosity	Gelation	Increase of viscosity	It greatly increases the viscosity (pseudogel)	It greatly increases the viscosity (pseudogel)	Gelation

Xanthan Gum

E-415

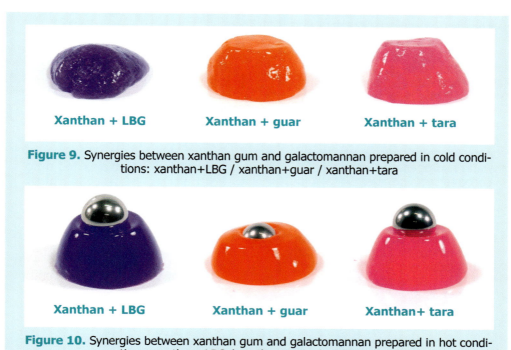

Figure 9. Synergies between xanthan gum and galactomannan prepared in cold conditions: xanthan+LBG / xanthan+guar / xanthan+tara

Figure 10. Synergies between xanthan gum and galactomannan prepared in hot conditions: xanthan+LBG / xanthan+guar / xanthan+tara

Examples of applications in food industry

- Emulsion stabilizer.
- Thickener in sauces and dressings.

FRESH SAUCE WITH MUSHROOMS

Ingredients: light cream, water, mushrooms 23,5% (mushroom, boletus: mixing in variable proportion of *boletus edulis, pinicola, aereus reticulatus*), cheese (Mascarpone, Grana Padano, Pecorino Romano cheese), whey powder, tapioca modified starch, sunflower oil, salt, margarine, acidity regulator (citric acid), preservative (potassium sorbate), spices, flavouring, thickener (**xanthan gum**).

- Beverage stabilizer.
- Ice cream mix stabilizer.
- Stabilizer in liquid food.

CONCENTRATE VEGETABLE BROTH

Ingredients: water, salt, vegetables (16%) (carrot, leek, red pepper, parsley, celery), vegetable fat, sugar, yeast, thickeners (**xanthan gum**, carob flour), flavourings (contains mustard), spices, caramel.

- Thickener in products like spreads and fillings, providing a feeling of creaminess.

HAM CROQUETTES

Ingredients: water, wheat flour, bread crumbs, vegetable oil, onion, whey powder, salt, margarine, flavouring, dry-cured ham (0,3%), spices, flavour enhancer (monosodium glutamate) and stabilizer (**xanthan gum**).

CREAMY CAKE WITH CHOCOLATE CREAM AND HAZELNUTS

Ingredients: sugar, vegetable oil, eggs, chocolate (sugar, cocoa paste, cocoa butter, defatted cocoa powder and flavouring), wheat flour, concentrated milk, glucose-fructose syrup, hazelnuts (4,3%), soya, stabilizer (glycerol), cocoa powder, egg white, hydrogenated vegetable fat, cocoa fat, raising agents (sodium bicarbonate, disodium diphosphate), thickeners (pectin, **xanthan gum**), salt, preservative (potassium sorbate), emulsifiers (soya lecithin, citric ester of mono- and diglycerides fatty acid), acidity regulator (citric acid), flavourings.

INSTANT CUSTARD PREPARED

Ingredients: sugar, dextrose, modified starch, stabilizers (diphosphate disodium, tetrasodium pyrophosphate, **xanthan gum**, sodium alginate, sodium phosphate), vegetable fat, milk protein, maltodextrin, acidity regulator (calcium acetate), emulsifiers (esters acetic of mono- and diglycerides of fatty acids, lactic esters of mono- and diglycerides of fatty acids), flavourings, salt and colour (tartrazine, sunset yellow).

- Thickener in baked and cooked doughs.

FISH SURIMI

Ingredients: white fish protein, water, vegetable oils, wheat flour, salt, vegetable protein, milk proteins, flavourings, flavor enhancer (monosodium glutamate), stabilizer (xanthan gum), acidity regulator (lactic acid), squid ink, garlic and chilli.

WHEAT TORTILLAS

Ingredients: wheat flour (61%), water, vegetable oil (fatty acid esters of ascorbic acid, tocopherol-rich extract), stabilizer (glycerin), salt, raising agent (sodium acid carbonate), emulsifier (mono and diglycerides of fatty acids), preservatives (potassium sorbate, calcium propionate), acidity regulators (malic acid and citric acid), thickener (**xanthan gum**), flour treatment agent (L-cysteine).

- Thickener in gluten-free bread doughs.
- Thickener in pet food (wet food).

10
E-416 KARAYA GUM

E-416 Karaya gum (CAS Number: 9000-36-6)

Other denominations: esterculia, tragacanth from India, kadaya, katilo, kullo, thapsi gum

Origin

Gum karaya is a hydrocolloid obtained from the exudate of trees of the genus *Sterculia*, mainly of *S.urens* that grows in India or *S.setigera*, found in Sudan (figure 1).

For years it was used to adulterate tragacanth gum due to their similar behavior and its cheaper cost.

The BIS (Bureau of Indian Standards) classifies the quality degrees of karaya gum according to the quality criteria shown in the table 1.

Composition and chemical structure

Karaya gum is a branched polysaccharide and partially acetylated (figure 2), formed by a main chain composed of galacturonic acid, rhamnose and galactose units. There are also small amounts of protein (Table 2).

An increase of galacturonic acid and methyl groups results in higher viscosity solutions, while an increase of galactose and arabinose decreases it.

Karaya gum loses its thickening capacity having been stored at room temperature. However, the effect is less noticeable when it is stored under refrigeration.

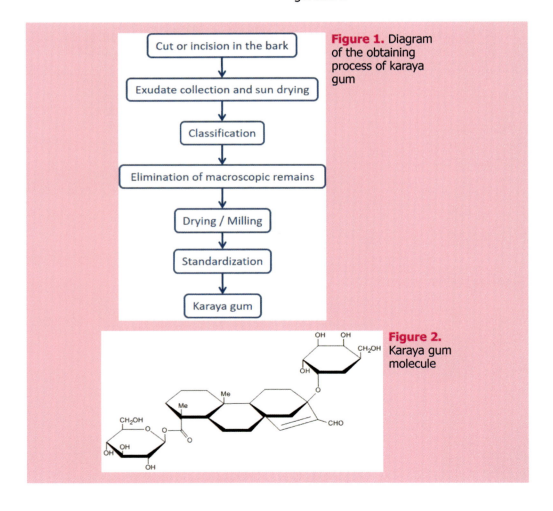

Figure 1. Diagram of the obtaining process of karaya gum

Figure 2. Karaya gum molecule

Karaya Gum

E-416

Table 1. Criteria of quality according to BIS

Grade	Color	BOFM(*)	Viscosity
HPS(**)	White	0,3-0,5 %	1200–1500 cps
I	White amber	0,5 %	1000 cps
II	Reddish yellow	1,5 %	600–1000 cps
III	Brown / Black	3 %	200–600 cps
Siftings - 1	Brown	6 %	
Siftings - 2	Dark brown	8–10 %	

(*)Bark and other foreign matter (**)Hand Picked Selected

Table 2. Composition of karaya gum

Proportion	Component
30-40%	Galacturonic acid
15-30%	Rhamnose
13-26%	Galactose
10-14%	Acetyl groups
4-5%	Glucuronic acid
1-2 %	Protein

Properties of solutions

Solubility and solutions preparation (Table 3)

Karaya gum is slightly soluble in cold water. Solutions must be heated above 80°C.

Karaya gum solutions are pseudoplastic, not very viscous and are slightly acidic (pH 5-6) for release of acetic acid.

Factors affecting the properties of solutions (Table 4)

- **Temperature:** karaya gum is not completely soluble in cold water, therefore, it is necessary to heat the solutions to get the correct solubility. Once properly solubilized, the solution viscosity is inversely proportional to temperature.

- **pH:** karaya gum is quite resistant to acidic medium, and the solubilization temperature decreases if the medium is acidified (pH 3-5).

Functionality

Karaya gum is a thickener that is insoluble in cold water, therefore it needs heating up to 80°C in order to solubilize (figure 3).

Table 3. Solubility and solutions preparation / +++ (high) ++ (middle) + (low)

Solubility in cold water	Solubilizing temperature	Premix	Shear force	Air incorporation	Increase viscosity of medium	Precipitates with alcohol
+	+++	+	+	No	++	Yes

Table 4. Factors affecting the properties of solutions

Factors	Effects on viscosity	Observations
Molecular weight / polymerization degree	Directly proportional	
Concentration	Directly proportional	
Temperature	Once solubilized, inversely proportional	An increase in temperature causes a decrease in viscosity
pH	Stable at acidic pH	The solutions are, in themselves, slightly acidic due to acetic release
Ionic charge	Does not affect	
Mechanical work	Pseudoplastic behaviour	Decreases viscosity as the shear force increases

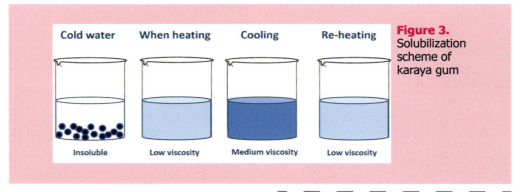

Figure 3. Solubilization scheme of karaya gum

Synergies or incompatibilities
They have not been described.

Examples of applications in food industry
- Maintains the viscosity in acidic ice creams and slows dripping.

LEMON-FLAVOURED ICE CREAM

Ingredients: milk, cream, water, skimmed milk, sugar, glucose syrup, whey, stabilizers (mono- and diglycerides of fatty acids, locust bean gum, guar gum, carrageenan), acidity regulator (citric acid), lemon flavouring, stabilizer (**karaya gum**), vanilla flavouring and colour (annato)

11 E-418 GELLAN GUM

E-418 Gellan gum (CAS Number: 71010-52-1)

Other denominations: gelana

Handbook of hydrocolloids

Origin

It is obtained by an aerobic fermentation of glucose by the bacteria *Sphingomonas elodea*.

Depending on the pH after pasteurization, products with different percentages of acetylation can be obtained, with different features on its functionality and use.

Composition and chemical structure

It is an anionic heteropolysaccharide, specifically a linear tetrasaccharide consisting of monomers of glucose, glucuronic acid and rhamnose with a 2:1:1 ratio.

According to its composition two types of gellan gum can be differentiated: high or low level of acetylation. High acetylation or high acyl gellan gum presents an acetyl group at de C6 of the glucose each 2 units of the tetrasaccharide (Figure 2). All of them have a glyceryl group at C2 of all glucose, and it has a spatial conformation of a levorotatory double-helix. Propiedades de las soluciones

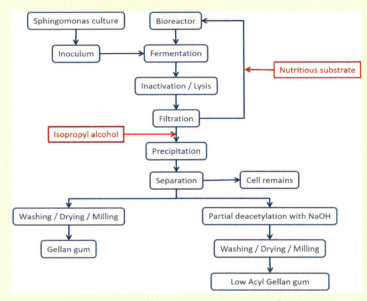

Figure 1. Diagram of the obtaining process of gellan gum

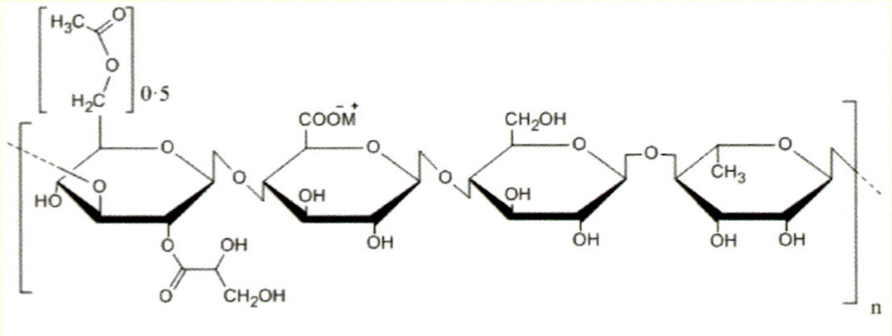

Figure 2. Gellan acetylated molecule

Gellan Gum

E-418

> This manual describes the characteristics of the low acetylation gellan or low acyl, since it is the one with a more widespread use in the food industry.

Properties of solutions

Solubility and solutions preparation

The gellan has no problems of dispersion, and the ions present in the medium may favor it. It does not require strong shear force or incorporates too much air, and it disperses easily in cold conditions, both water and milk.

In general, the complete solubility of the gellan occurs around 80-85ºC, it requires a small amount of calcium sequestrant as phosphates and citrates, and it is stable at acidic pH.

Factors affecting the properties of solutions

- **Ionic charge:** the solubility of gellan gum is clearly conditioned by the ionic charge of the medium, since it increases the necessary temperature to solubilize the gellan and may even inhibit the solubilization depending on the concentration of divalent ions, such as calcium or magnesium.

Functionality

Gelation mechanism

The most commonly used method of gellan gelation occurs after a correct solubilization at temperatures close to boiling and the subsequent addition of a monovalent (K^+, Na^+) or divalent cation (Mg^{2+}, Ca^{2+}) once the gum is solubilized.

As in the agar case, gellan presents a certain hysteresis between solubility and gelling temperatures. In this case, the gelation occurs when there is a drop in the temperature up to 50-70ºC, depending on the amount of ions in the medium. In Figure 3 you can see as the concentration of ions Ca^{2+} in the medium increases, the starting temperature of gelation also elevates.

The gelation of gellan without adding monovalent or divalent ions is also possible by decreasing the pH of the medium. To get a gellan gel by using this method, the dispersion of the gellan must be in a medium at pH>3, and this pH must be adjusted before the solubilization of the gellan by temperature. The gelation is also viable if solubilization occurs before adjusting the pH to below 3, although the formed gels will have less strength, as shown in Figure 5.

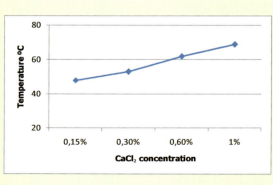

Figure 3. Graph of temperatures of gellan gelation with different concentrations of Ca^{2+}

Table 1. Solubility and solutions preparation / +++ (high) ++ (middle) + (low)

Solubility in cold water	Solubilization temperature	Premix	Shear force	Air incorporation	Increase viscosity of medium	Precipitates with alcohol	Ion sequestrant
No	+++	+	+	+	+	No	++

Table 2. Factors affecting the properties of solutions

Factors	Effects on viscosity	Observations
Molecular weight / polymerization degree	Directly proportional	
Concentration	Directly proportional	
Temperature	Directly proportional	An increase in temperature causes a slight increase in the viscosity
pH	It is stable over a wide pH range	If solubilization occurs at pH lower than 3, it will gel during cooling
Ionic charge	It decreases the viscosity	Divalent ions may inhibit gelation since it prevents solubilization
Mechanical work	Pseudoplastic behaviour	It decreases the viscosity as it increases the shear force

Gellan Gum

E-418

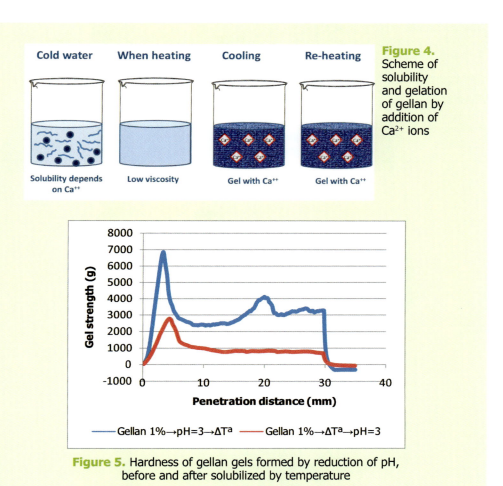

Figure 4. Scheme of solubility and gelation of gellan by addition of Ca²⁺ ions

Figure 5. Hardness of gellan gels formed by reduction of pH, before and after solubilized by temperature

Characteristics of gels

Formed gels are different depending on the type of gellan used. Those of low acetylation (Figure 6), the most commonly used, result in a firm, rigid and brittle gel presenting syneresis in cutting. Instead, with high acyl gellan, elastic and soft gels are obtained and, generally, do not present syneresis.

Low acyl gels are thermostable if they are formed by addition of divalent ions after solubilization. On the other hand, they are thermorreversibles and have a very noticeable hysteresis loop, similar to the agar, when they are formed by addition of monovalent ions, especially potassium and in a diary medium, by Ca^{2+} ions. The high acetylation ones are thermorreversibles but do not have hysteresis, they melt almost at the same temperature that set.

Gellan gum, as well as being gelling agent, also has the characteristic of forming a "fluid gel" at concentrations much lower than 1%, which allows the suspension of large-size particles. This effect is achieved by applying a shear force to the product during or after gelation. The result is a very weak, liquid gel that flows without problems, but is able to keep in suspension large particles, as you can see in Figure 7, which represents suspended spheres of sodium alginate in a "fluid gel" gellan.

Handbook of hydrocolloids

Table 3. Characteristics of gels

Type of gellan	Low acetylation	High acetylation
Gel type	Firm and crisp	Soft and elastic
Syneresis	Yes, only in the cut	No
Temperature stability	Thermostable in the presence of Ca^{2+}. Thermorreversible with K^+ and in milk, hysteresis	Thermorreversible, no hysteresis

Figure 6. Low acetylation gellan gel

Figure 7. Alginate spheres suspended in a gellan "fluid gel" at 0,05%

Synergies or incompatibilities

- Gellan gum + gelatin: the use of low acyl gellan gum with gelatin increases the set temperature of gelatin. It also increases its melting temperature, so gelatin-based products can withstand more time at room temperature without losing their shape.

Examples of applications in food industry

- Gelling agents in jams with low concentrations of soluble solids (low sugar).

- Increase stability of gelatinous desserts like jellies taking advantage of the existing synergy between the two products.

- Film formation to reduce the absorption of frying oils.

- Formation of a "fluid gel" to suspend particles in drinks.

COCONUT BEVERAGE WITH PULP

Ingredients: mineral water, coconut cream, sugar cane, coconut juice, stabilizer (**calcium lactate**), coconut natural flavouring, acidifier (citric acid), antioxidant (vitamin C), stabilizers (sodium citrate, **gellan gum**), grapefruit seed extract.

SOYA DRINK

Ingredients: water, soya beans (10,3%), sugar, stabilizers (**tricalcium phosphate, gellan gum**), flavouring, salt and vitamin B12.

12 — E-425 KONJAC

E-425 Konjac (CAS Number: 37220-17-0)

Other denominations: konjac flour, konjac manano, konnyaku, glucomannan.

Handbook of hydrocolloids

Origin

Konjac is obtained from the tuber *Amorphophallus konjac* containing around a 60-80% of konjac, by grinding it either dry or wet. Its use is native to China and Japan.

Composition and chemical structure

It is a non-ionic heteropolysaccharide with a high molecular weight compound with glucose and mannose connected by β(1-4) links with a ratio of 3:2. The main chain is linear with acetylated groups every 9-19 glucose/mannose units. These acetylated groups contribute to the solubilization of the product.

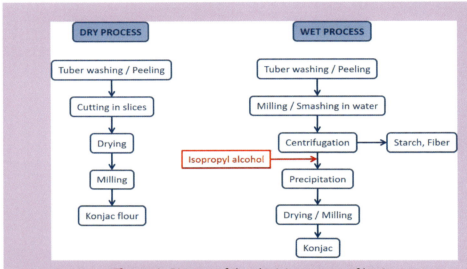

Figure 1. Diagram of the obtaining process of konjac

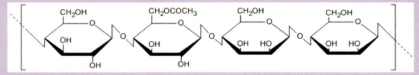

Figure 2. Konjac molecule

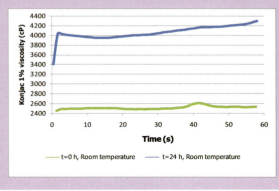

Figure 3. Graph of viscosity as a function of the time. Rotational viscometer (50 rpm, 60 s, spindle R4 and R5)

Konjac
E-425

> In Figure 3, two different types of spindles were used, since there was too much difference in viscosity between samples. The spindle R4 was used to measure the viscosity of the sample at time 0, but was out of range for the same sample after 24h. Therefore, it was necessary to use a higher spindle, R5.

As the gum moisturizes, it forms a solution of high viscosity with pseudoplastic behavior, which precipitates with alcohol, is stable over a wide pH range and is not affected by a change in ionic charge.

Factors affecting the properties of solutions

- **Time:** konjac increases its viscosity as time passes. (Figure 3). These variables were checked with a dissolution to 1% of konjac in which the viscosity was measured in a rotational viscometer, at room temperature (25°C), in a freshly prepared sample (time 0) and that same sample after 24 h had passed.

- **pH:** konjac is a hydrocolloid that is stable in a wide range of pH, 3 to 10. Below 3 or above 10, konjac loses its viscosity, but does not precipitate.

Properties of solutions

Solubility and solutions preparation

Konjac is soluble in cold water with a slow hydration rate, allowing it to disperse without using shear force; this distinguishes it from other gums such as guar or xanthan which rapidly hydrate.

Table 1. Solubility and solutions preparation / +++ (high) ++ (middle) + (low)

Solubility in cold water	Solubilization temperature	Premix	Shear force	Air incorporation	Increase viscosity of medium	Precipitates with alcohol
+++	Cold	+	+	No	+++	+++

Table 2. Factors affecting the properties of solutions

Factors	Effects on viscosity	Observations
Molecular weight / polymerization degree	Directly proportional	
Concentration	Directly proportional	
Temperature	Indirectly proportional	An increase in temperature causes a decrease in viscosity
pH	Stable over a wide range (3-10)	
Ionic charge	Does not affect	
Mechanical work	Pseudoplastic behaviour	Decreases viscosity as the shear force increases

Handbook of hydrocolloids

Functionality

Gelation mechanism

Once the konjac is solubilized, a heating and a subsequent cooling stage is needed to produce the gelation. This hydrocolloid, at the conditions to which it is subjected, has the characteristic that can form two types of gels:

- **Thermoreversible gels:** they are formed by synergy with other gums such as kappa carrageenan, agar or xanthan gum. By itself, at a concentration of 10%, it is a very viscous pseudogel.

- **Thermostable gels:** are formed by deacetylation of konjac by a weak base such as potassium carbonate.

Form of preparation

- **Thermoreversible gels:** dispersion of the two gums in water, heating stage above 85ºC and the subsequent cooling stage.

- **Thermostable gels:** solubilize and hydrate well the konjac, get to a temperature above 85ºC and hold for 30 minutes with continuous agitation. Stop agitation and leave to cool down to room temperature. Then add the weak base (for example, K_2CO_3) until reaching a pH of 9 or higher to deacetylate the molecule and mix well. Begin to heat the mixture again to 85ºC for 2h without agitation. Leave to cool.

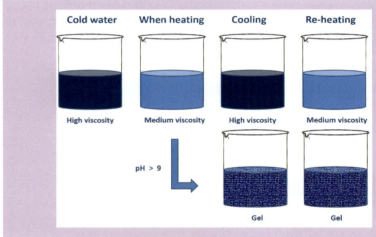

Figure 4. Solubilization and gelation scheme of konjac

Depending on if we talk about flour konjac or konjac, the purity and the functionalities described may vary. Thus, the konjac flour will have more impurities, having to add more to get the same result as that obtained with pure konjac.

In the bibliography, it can be found that this type of gel is formed to minimal concentrations of 1,5-2% of konjac; this test has been made in the BDN lab, and the minimum concentration needed to form the gel is 5%.

Characteristics of gels

Two types of gel can be formed according to the factors present in the medium and its features can also be different, as it shows table 3.

The gel-forming ability of konjac gives its capacity of forming films, which have a great strength and cohesiveness.

Factors affecting the properties of gels

- **Temperature:** it must reach temperatures higher than 85ºC.

- **Presence of a weak base:** by deacetylating the konjac molecule, it will lose its solubility, creating a three-dimensional network by the formation of hydrogen bridges that altogether configured the gel.

- **Presence of a humectant** (like glycerin): by increasing the % moisturize, the molecule loses strength and endurance of the film but increases the adhesion.

Synergies or incompatibilities

- Konjac + kappa carrageenan: see carrageenan synergies (figure 5)

- Konjac + xanthan gum: this synergy leads to a great increase of viscosity and/or a gelation.

On one hand, a mixture of konjac (0,8%) with xanthan gum (0,2%) makes the viscosity increase up to 3 times by comparing products on their own 1%.

However, on the other hand, the formed gels are thermoreversible, cohesive, elastic and very stable at pH. Its maximum synergistic effect is at pH 5. The formed gel has characteristic viscoelastic properties since it is able to endure much pressure without breaking and regains the initial shape when pressure is not being applied. Maximum gel strength found is in the mixture of 0,4% of konjac and 0,6% of xanthan gum (by themselves, both the

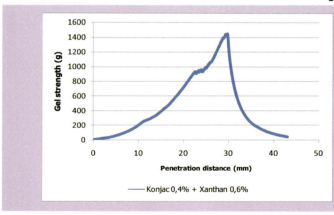

Figure 6. Gel strength graph of the synergy konjac-xanthan gum

Table 3. Characteristics of gels

	Konjac	Konjac + xanthan	Konjac + kappa	Konjac+agar
Gel type	Elastic and firm gel	Weak and elastic gel	Elastic and firm gel	A little brittle and elastic gel
Syneresis	Yes	Yes	No	Yes
Temperature stability	Thermostable at pH>9	Thermoreversible	Thermoreversible	Thermoreversible

Handbook of hydrocolloids

konjac and the xanthan gum do not gel).

- Konjac + Agar: this synergy gives a thermoreversible gel with a maximum gel strength when mixing 0,2% of konjac with 0,8% of agar. As shown in Figure 9, the hardness of the agar is increased about 10 times after the addition of a small amount of konjac.

- Konjac + starches: the synergy decreases the syneresis and increases the viscosity of the solution, which is maintained during cooking and subsequent cooling. This feature is very useful in low-calorie products because it decreases the amount of added starch and improves texture. Synergy was 1:5 potato:starch ratio.

Figure 5. Kappa carrageenan and konjac gel

Figure 7. Konjac and xanthan gum gel

Figure 8. Konjac and agar gel

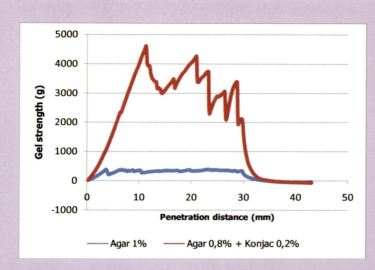

Figura 9. Comparative graph of gel strength of agar gel versus the synergy with konjac

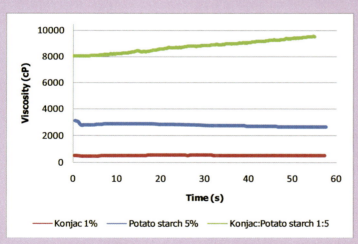

Figure 10. Viscosity graph of the synergy of konjac and potato starch. Rotational viscometer (100 rpm-250 rpm, 60 s, R6 and R7 spindle at temperature 25ºC)

As in the case of the previous graph, two types of spindle were used due to the great difference in viscosity between samples. For the same reason, the rotation speed is varied.

Examples of applications in food industry

- Thickener in pastry, dairy products (yogurt, puddings, ice creams), jams, jellies and beverages

GRAPE JUICE

Ingredients: grape juice (25%), fructose syrup, stabilizer (**konjac**).

Handbook of hydrocolloids

- Gelling agent in baking products, jams, jellies, meat products (ham), candy, by its own or in synergy with other hydrocolloids.
- Fat replacer in meat products.
- Reduces syneresis in frozen products.

TURKEY AND CHEESE ROLL

Ingredients: Covering (50%) (wheat flour, water, potato starch, sunflower oil, salt, sugar, wheat gluten, yeast), filling (50%) (tomato sauce (34%), water, sunflower oil, salt, sugar, spices, herbs, turkey meat prepared (turkey (26%), water, potato starch, salt, gelling agents (carrageenan and **konjac**), glucose syrup, dextrose, stabilizers (sodium triphosphate and sodium diphosphate), vegetable fiber, flavour enhancer (monosodium glutamate), soy protein, flavorings, antioxidant (sodium erythorbate), preservative (sodium nitrite), mozzarella cheese (20%), Edam cheese (20 %)).

> ℹ️ Konjac, for its composition, is a source of soluble dietary fiber from low-calorie feature used in low energy products.

COCONUT JELLY

Ingredients: sugar, fructose, coconut, gelling agent (**konjac**), algae extract, acidity regulator (citric acid), flavouring.

- Stabilizer in bakery and dairy products (yogurt, ice cream).
- Water retainer in bakery and prepared meat.
- Stabilizer during cooking in bakery.
- Provides fiber in bakery (bread).
- Retards recrystallization in ice cream.
- Film-forming in edible films.

Konjac

E-425

 Killer candies

In 2002 the European Union suspended the marketing and import of jelly confectionery containing konjac. This suspension is collected in the *COMMISSION DECISION of 27 March 2002 suspending the placing on the market and import of jelly confectionery containing the food additive E 425 konjac (2002/247/EC)* concluding that the microcapsules containing konjac constitute a risk to the health and life of persons, based on the consumption of these sweets, which had caused the death by drowning of several children and elderly people in the United States, Canada and United Kingdom. This restriction remains valid in the current regulations (*Regulation (EU) No 1129 / 2011 from the Commission of 11 November 2011 amending Annex II to Regulation (EC) no 1333 / 2008 of the European Parliament and of the Council to establish a list of food additives of the Union*), but it can be used in other applications.

13 E-440 PECTIN

E- 440 Pectin (CAS Number: 9000-69-5)

Other denominations: apple gelatin.

Origin

Pectin is extracted from the cell walls of certain vegetables like the skin of citrus fruits, apples, beet or sunflower (figure 1). On the cell wall of these vegetables is the protopectin, an insoluble form of pectin when the fruit is ripe, as its maturing process produces the natural methylation of the polymer, which is known as pectin.

It is crucial in the inactivation of enzymes during the extraction process, since the presence of pectinesterases and poligalacturonases causes the depolymerization and demethylation of the molecule, inactivating its gelling capacity.

Depending on the conditions applied during the extraction process we can obtain high methoxyl pectin (HM pectin), low methoxyl pectin (LM pectin) pectin, or amidated low methoxyl pectin (ALM pectin).

Composition and chemical structure

Pectin is an heteropolysaccharide consisting mainly of a linear chain of galacturonic acid linked by α(1-4) bonds. Some of the carboxyl radicals in the molecule are esterified with methoxy groups (also called methyl groups) (figure 2). Depending on the percentage of esterified methyl groups, the following classification of pectin is set:

- **HM pectin**. Also called high metoxil pectin, HE pectin or high ester pectin. There are more than 50% of the carboxyl groups -COOH esterified with methyl groups -CH$_3$. Commercially, they are most frequently those who have between 60% to 80% esterified groups.

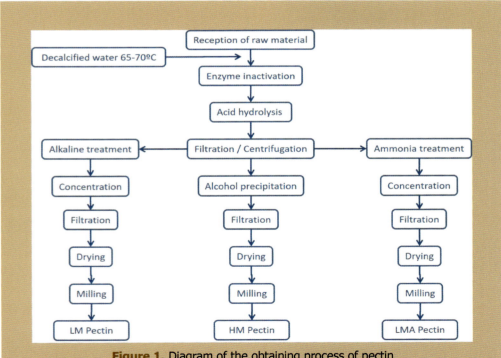

Figure 1. Diagram of the obtaining process of pectin

- **LM pectin.** Also called low metoxil pectin, LE pectin or low ester pectin. There are less than 50% of the carboxyl groups -COOH esterified with methyl groups - CH_3. Commercially, they are most frequently those who have between 20 to 40% of esterified groups.

- **ALM pectin.** They are LM pectin, low methoxyl pectin, which also have been esterified with amide groups by reacting with ammonia. The degree of amidation (DA) is the percentage of amide groups associated with the main chain of galacturonic acid of the pectin molecule. The level of amidation in the carboxyl groups cannot be greater than 25%, according to the *COMMISSION REGULATION (EU) No 231/2012 of 9 March 2012 laying down specifications for food additives listed in Annexes II and III to Regulation (EC) No 1333/2008 of the European Parliament and of the Council.*

> ⓘ Although LM pectin generally works better after it is solubilized in hot conditions, it can give some reactivity without having gone through a stage of warming and lead to gelling in the presence of divalent ions (Ca^{2+} and others).

Properties of solutions

Solubility and solutions preparation (table 1)

Pectin is partially soluble in cold water, obtaining greater functionality when it solubilizes completely in hot conditions.

Pectin hydrated in water provides certain viscosity that will vary depending on the concentration of pectin and the degree of polymerization of the molecule. The gelation of the pectin, even with HM or LM pectin, can only take place with certain factors in the medium.

LM pectin is reactive in the presence of divalent ions in the medium, so to ensure their proper hydration it is necessary for it to take place in the absence of these ions. Therefore, LM pectin does not start gelling during the stage of agitation.

Factors affecting the properties of the solution (table 2)

- **Temperature:** it facilitates the pectin solubilization. In the case of HM pectin, it is necessary to heat the solution, therefore it solubilizes. In the case of LM pectin, it can also work hydrated in cold conditions.

- **Ionic charge:** in the case of LM pectin, they have to hydrate in the absence of divalent ions in the medium. If you need, sequestrants (phosphates or others) will be available to use.

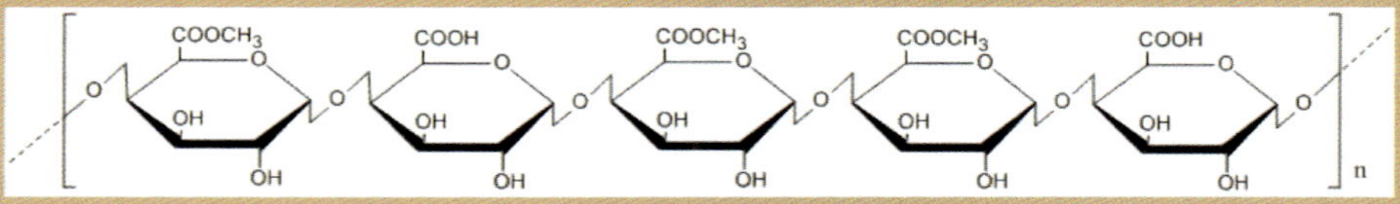

Figure 2. Pectin molecule

Handbook of hydrocolloids

Table 1. Solubility and solutions preparation / +++ (high) ++ (middle) + (low)

Pectin	Solubility in cold water	Solubilizing temperature	Premix	Shear Force	Air incorporation	Increase viscosity of medium	Precipitates with alcohol	Ion sequestrant
HM Pectin	+	+++	+	+	No	++	No	No
LM & LMA pectin	++	+++	+	+	No	++	No	Yes

Table 2. Factors affecting the properties of solutions

Factors	Effects on viscosity		Observations
	HM Pectin	LM Pectin	
Molecular weight/ polymerization degree	Directly proportional	Directly proportional	
Concentration	Directly proportional	Directly proportional	
Temperature	Inversely proportional	Inversely proportional	An increase in temperature causes a decrease in viscosity
pH	It is stable over a wide range of pH	Does not affect	
Ionic charge	Does not affect	The presence of divalent ions cause an increase of viscosity	Avoid divalent ions by using sequestrant
Mechanical work	Pseudoplastic behaviour	Pseudoplastic behaviour	Viscosity decreases as it increases the shear force

Functionality

Gelation mechanism

Pectin solutions make gel in specific conditions of the medium. These gelation factors vary depending on the type of pectin you will work with:

- **HM pectin:** they make gels with high concentrations (≥ 55%) of soluble solids and an acidic pH (between 3,5-3,8). The most used soluble solids are sugars, although all ingredients that are soluble in water at a concentration equal to or greater than 55%, would be useful to form HM pectin gel, along with and acidic pH.

- **LM pectin:** they make gels in the presence of divalent ions in the medium. The most commonly used are calcium salts, like calcium lactate, calcium chloride, calcium sulphate, etc. (see characteristics of solubility of these calcium salts in table 3 of chapter 3)

- **ALM pectin:** they work the same as LM pectin, however, it does not need an extra intake of calcium in the medium from an external calcium salt, in fact, with the amount of calcium that may exist in the medium, it is able to react and form a gel.

- **Gelation conditions of HM pectin:** to favor the attraction of pectin molecules and form a gel, a high concentration of sugar is needed until it enters competition through the water with the pectin. By reducing the attraction of pectin by water, since the water is retained by the sugar, they are established unions of pectin, connecting molecule to molecule forming a gel that immobilizes the water in the form of syrup by the high concentration of sugar. The addition of acid to 3,5 pH values brings ions H^+ (hydrogen ions) to the medium and, as a result, free –COOH groups of pectin decrease their ionization, reducing its electrical charge and favoring its union to form the gel (figure 5).

- **Gelation conditions of LM pectin:** gel formation with LM pectin is similar to that which is established in the case of alginate (egg box structure). Calcium ions create a link between two face-to-face strands of pectin, by that forming the gel (figure 7).

Factors affecting the properties of gels

Depending on the type of pectin used to form the gel, a series of factors have to be taken into consideration, which are described in tables 3 and 4.

> **i** Intermediate pectin, i.e., those that are not either HM or LM, and whose levels of esterified methyl groups are around 50%, respond to intermediate reaction mechanisms. It is correct to say that they need sugar, acid and calcium ions to gel.

> **i** HM and LM pectin, hydrated in a medium with a 37% of alcohol, produce a gelation without needing the classic factors of gelling in each of the cases (HM: acid pH + 55% soluble solids. LM: calcium ions)

Handbook of hydrocolloids

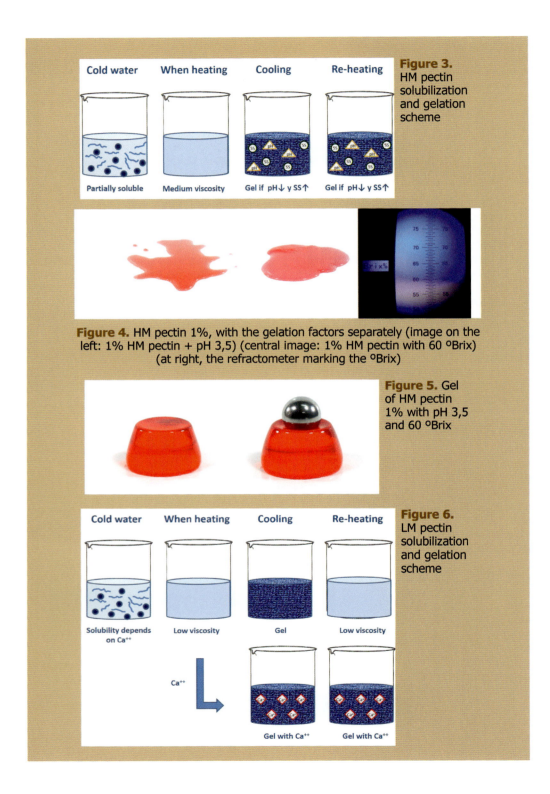

Figure 3. HM pectin solubilization and gelation scheme

Figure 4. HM pectin 1%, with the gelation factors separately (image on the left: 1% HM pectin + pH 3,5) (central image: 1% HM pectin with 60 °Brix) (at right, the refractometer marking the °Brix)

Figure 5. Gel of HM pectin 1% with pH 3,5 and 60 °Brix

Figure 6. LM pectin solubilization and gelation scheme

Figure 7. Reverse gelation with LM pectin

Table 3. Factors affecting the properties of HM pectin gels

\	HM pectin	\
Factors	**Effects on gelation**	**Observations**
Degree of esterification (DE) of pectin	A higher percentage of methyl groups esterified, faster gelation	Rapid set (DE> 65%) Slow set (DE< 65%)
Temperature	Required initially to solubilize. Once solubilized, inversely proportional. Different set temperature according to type of pectin	Thermostable Rapid Set (85-75ºC) Medium set (75-60ºC) Slow set (60-50ºC)
Soluble solids	Essential a ≥55% concentration to make a gel	If that concentration is not achieved, it will not gel
Shear force	Gel break	Syneresis
Concentration	Directly proportional	At higher concentration, more gel strength
pH	Needs a pH between 3,5-3,8 to make a gel	If that pH is not achieved, it will not gel

Table 4. Factors affecting the properties of LM pectin gels

\	LM pectin	\
Factors	**Effects on gelation**	**Observations**
Temperature	Required initially to solubilize. Once solubilized, inversely proportional	Thermoreversible at temperature > 100ºC
Ionic charge	Low or not soluble	
Shear force	It breaks the gel	No syneresis
Concentration	Directly proportional	At higher concentration, more gel strength
Presence of divalent ions	Necessary for gelation	Concentration determines the hardness

Table 5. Characteristics of gels

Type of pectin	HM pectin	LM pectin	LMA pectin
Type of gel at pH<3,5	Rigid gel	Spreadable soft gel	Spreadable soft gel
Type of gel at pH>3,5	It does not gel Viscosity	Spreadable gel Thixotropy	Spreadable gel Thixotropy
Syneresis	Yes	No	No
Temperature stability	No	Yes, at T >100ºC	Yes, at T <100ºC
Shear force	It breaks the gel	Reversible break according to the pH	Reversible break at pH> 3,4

> HM pectin are hydrated and solubilized in hot conditions reaching temperatures of 90-95ºC. Thus, during the cooling stage, the gelation starts. The temperature in which the gel starts to form is called as "set temperature" and includes a range of temperatures in which the solution is ordered spatially giving a solid (gel).
>
> HM pectin are classified among "RAPID SET", "MEDIUM SET" or "SLOW SET" pectin, depending on the starting temperature of gelation, which is related in turn with the starting speed of gelation. This difference in the starting temperature of gelation is due to the degree of esterification of pectin with methyl groups. The greater the percentage of esterification is, the faster it reacts and, therefore the sooner it gels.
>
> -RAPID SET: gelation speed is fast since it starts to gel at a temperature around 85-75ºC and a esterification degree around 70-76%
>
> -MEDIUM SET: gelation speed is intermediate since it starts to gel at a temperature around 75-60ºC and a esterification degree around 68-70%
>
> -SLOW SET: gelation speed is considered slow since it starts to gel at a temperature around 60-50ºC and a esterification degree around 60-67%

Characteristics of gels

Depending on the type of pectin that is used, (HM pectin, LM pectin and ALM pectin) the solution produces a gelation with different characteristics, shown in table 5.

Synergies or incompatibilities

- Interaction of HM pectin with proteins: HM pectin has a specific, useful application in the stability of beverages with acidic pH, with the presence of proteins. It is the case of drinks that combine milk or soy-based drinks with fruit juice, etc. The proteins in colloidal suspension in a liquid medium (when there is an acidic pH close to the isolectric point of the protein), precipitate as they have a total net charge of 0, and lose the strength of repulsion between them that was allowing them to stay in suspension without forming a precipitate.
It is possible to avoid the unwanted effect of precipitation of the protein if, prior to adding acid to the colloidal protein system (milk or others), we disperse and hydrate HM pectin. This is because hydrated HM pectins provide negative charges in the medium, which will act as a "cushion effect" that will prevent protein sediments (figure 8).

- Pectin + sodium alginate: they can form thermoreversible gels without the presence of calcium ions, in acidic conditions.

Pectin

E-440

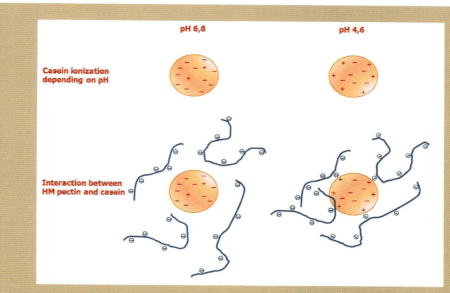

Figure 8. Synergistic behavior between HM pectin and the protein at its isoelectric point

Examples of applications in food industry

- Gelling agent in jams and fillers.

RASPBERRY FRUIT PASTE

Ingredients: fruit pulp 50% (raspberry 27%, apricot 19%, pear 4%), sugar, glucose syrup from wheat, invert sugar, wheat germ 6%, stabilizer (sorbitol), natural flavouring, gelling agent (**pectin**), acidity regulator (citric acid), antioxidant (ascorbic acid), vitamin B1.

- Beverage stabilizers.

FERMENTED MILK DRINK

Ingredients: skimmed milk, inulin, dextrose, stabilizer (**pectin**), lactic ferments (yoghurt ferments and Lactobacillus casei), sweeteners (aspartame and acesulfame-K), flavouring.

FRUIT JUICE DRINK

Ingredients: water, pulp of soursop (18%), juice grape (3%), sugar, acidity regulator (citric acid), stabilizer (pectin), flavourings.

- Thickening, gelling agent in spreadable texture products.

LIGHT BUTTER

Ingredients: butter, water, emulsifiers (mono and diglycerides of fatty acids, citric esters of monoand diglycerides of fatty acids, polyricinoleate of polyglycerol), natural flavouring, sea salt, acidity regulator (lactic acid), stabilizer (**pectin**), preservative (potassium sorbate), colour (beta-carotene), vitamin A and vitamin E.

- Give gelled texture in candies and jellybeans.

Handbook of hydrocolloids

- Give shine to confectionery and bakery coats and other aesthetic applications.

- Gelling agent in desserts of gelled water (jelly).

- Thickener in yogurt smoothies.

AQUEOUS DESSERT WITH GELATIN AND STRAWBERRY PULP

Ingredients: water, sugar, glucose and fructose syrup, strawberries pulp (4%), gelatin (2,6%) acidity regulators (citric acid and sodium citrate), gelling agent (**pectin**), flavourings, preservative (potassium sorbate), vitamin C and colour (allura red AC).

PASTEURIZED PREPARED MILK (STRAWBERRY)

Ingredients: semi-skimmed yogurt pasteurized after fermentation (40%), water, sugar, strawberry juice (5,2%), dextrose, stabilizer (**pectin**), salts of calcium, phosphorus and magnesium, flavoring and colour (cochineal).

> Pectins, regardless of the type, are universally better accepted by their denomination than the E number that identifies them as a food additive, and so it is often found in the list of ingredients identified as "pectin" instead of the number E-440.
>
> Even though there are HM, LM and ALM pectin, they share a unique E number, the E-440 (as it happens in the case of the carrageenan).

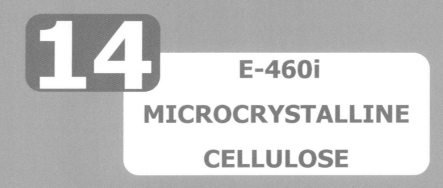

14 E-460i MICROCRYSTALLINE CELLULOSE

E-460i Microcrystalline Celulose or cellulose gel (CAS Number: 9004-34-6)

Other denominations: MCC.

Handbook of hydrocolloids

Origin

Cellulose is a molecule formed by glucose units connected by β (1-4) links, which is located in the cellular walls of the plants. In particular, MCC is obtained from the cellulosic fraction of purified wood and cotton pulp, partially depolymerized using a controlled acid hydrolysis.

> With the publication of the Regulation 1129/2011 amending Annex II to Regulation 1333/2008, the denomination microcrystalline cellulose disappeared because of the disappearance of the subsections i), to be called simply cellulose. However, in regulations of later publications as Regulation 231/2012 about specifications for food additives or Regulation 1274/2013 amending annexes II and III of the 1333/2008, reappears with the name "E-460i cellulose microcrystalline" and it is also added the denomination "cellulose gel".

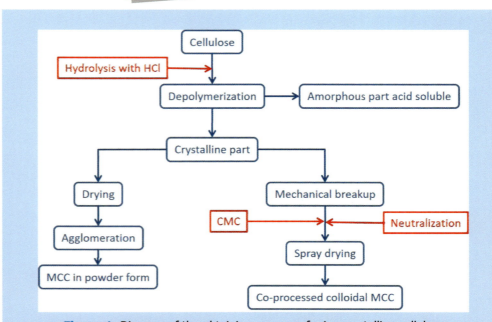

Figure 1. Diagram of the obtaining process of microcrystalline cellulose

Composition and chemical structure

Microcrystalline cellulose is a linear non-ionic monopolysaccharide. It retains the same chemical structure as the native cellulose, although it only consists of the insoluble fraction, called α-cellulose.

The types of MCC are available according to its composition:

- MCC powder grade: is a 100% α-cellulose, which gives rise to a completely inert and insoluble product.

- MCC colloidal grade: composed of 75-90% for α-cellulose and a 10-25% for a hydrocolloid that acts as a hydrophilic agent and favors the dispersion of the MCC. Usually this hydrocolloid is carboxymethylcellulose or CMC, a ionic ether of cellulose, although xanthan gum, guar gum, and more can also be used.

Properties of solutions

Microcrystalline cellulose is completely insoluble it is only dispersible. The colloidal grade MCC will disperse more easily than the powder grade due to the action of hydrocolloids which are part of its composition.

Factors affecting the properties of dispersions

Despite being insoluble, the dispersion of microcrystalline cellulose in the medium can be affected by several factors. In the case of the MCC colloidal grade, the drying system that is used in the procedure of obtaining the hydrocolloid will be decisive for its ease of dispersion. Those celluloses obtained by drying by atomization will be dispersible simply through mechanical agitation, while the ones that are dried by fluidized bed will need a stage of homogenization for a correct dispersion.

Another factor to take into account when making the dispersion is the amount of minerals present in the medium, for example milk, in which case, the calcium present can hinder the dispersion and also require the application of an homogenization stage.

Functionality

The main functionality of microcrystalline cellulose, in the case of the MCC colloidal grade, is the formation of a 3D network that provides the medium the characteristics of a "fluid gel", meaning a structure capable of suspending particles of a certain size, as for example powder cocoa in drinks, keeping a completely fluid medium.

In addition to helping suspend particles of large size, this invisible 3D network which forms the MCC colloidal grade helps the stabilization of emulsions and foams since it avoids the coalescence of fat droplets. It also helps to create a fat mouthfeel, so it can be used as a fat replacer in products with a reduced fat content.

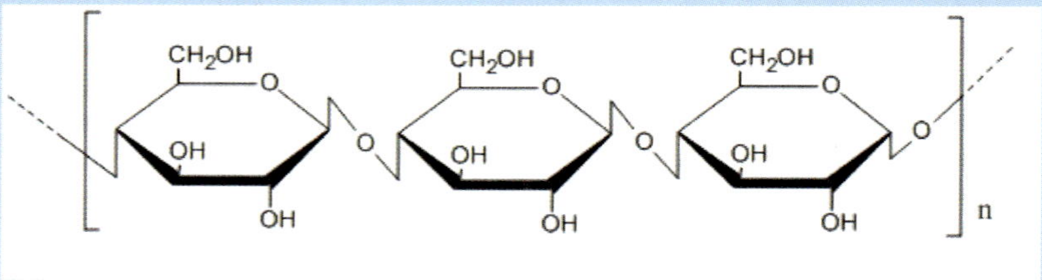

Figure 2. Microcrystalline cellulose molecule

Table 1. Solubility and preparation of the solutions / +++ (high) ++ (middle) + (low)

MCC	Solubility in cold water	Solubilization temperature	Premix	Shear force	Air incorporation	Increase viscosity of medium	Precipitates with alcohol	Ion sequestrant
Grade powder	Insoluble	Insoluble	+++	+	No	No	No	No
Colloidal grade	++	+	++	+++	No	+	No	+

Table 2. Factors affecting the properties of dispersions

Factors	Effects on viscosity		Observations
	Powder grade	Colloidal grade	
Molecular weight / polymerization degree	Does not affect	Directly proportional	
Concentration		Directly proportional	
Temperature	Does not affect	Does not affect	The temperature has low effect on the functionality
pH	Does not affect	It is stable over a wide pH range	
Ionic charge	Does not affect	It decreases the viscosity	Ions such as calcium may hind the dispersion
Mechanical work	Not soluble	Pseudoplastic by action of the CMC	It decreases the viscosity as it increases the shear force

Microcrystalline cellulose

E-460i

In the case of the MCC powder grade, which is completely inert and insoluble, its functionality is clearly linked to its use as an anti-caking agent, or bulking agent of other substances. Its use can also be considered as dietary fiber since it is not absorbed by the digestive tract.

Synergies or incompatibilities

There are no synergies described between the MCC and other hydrocolloids, apart from the already discussed with CMC (MCC colloidal grade).

Examples of applications in food industry

- Suspension of particles in cocoa beverages, after formation of 3D network in combination with kappa carrageenan

---
UHT COCOA DRINK

Ingredients: whole milk, skimmed milk, whey, sugar, defatted cocoa (1,5%), stabilizers (**microcrystalline cellulose, carboxy methyl cellulose**, sodium phosphates, carrageenan) and flavouring

---

- Stability of the fat phase in light cream

UHT LIGHT CREAM 12% F.M.

Ingredients: 12% F.M. cream, stabilizers (carrageenan, **microcrystalline cellulose and carboxy methyl cellulose**)

- Anti-caking agent in grated cheese.

MIXTURE OF GRATED CHEESE

Ingredients: mozzarella (59%) (pasteurized milk, salt, curdling agent, ferments), emmental (39%) (pasteurized milk, ferments, curdling agent, salt), anti-caking agent (**microcrystalline cellulose**)

- Stability in vegetable whipped cream

U.H.T. VEGETABLE WHIPPED CREAM

Ingredients: water, vegetable fat, sugar, casein, stabilizers (sorbitol, **carboxy methyl cellulose, microcrystalline cellulose**), emulsifiers (lactic esters of mono and diglycerides of fatty acids, lecithin), salt, sequestrants (sodium phosphate, sodium citrate), colour (carotenes) and flavourings

---

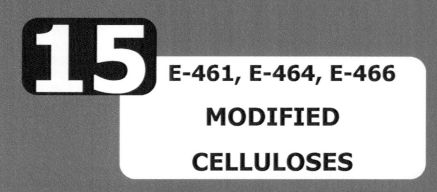

E-461 Methylcellulose (CAS Number: 9004-67-5)

Other denominations: cellulose methyl ether, MC

E-464 Hydroxypropyl methyl cellulose (CAS Number: 9004-65-3)

Other denominations: hypromellose, HPMC

E-466 Carboxy methyl cellulose or celulose gum (CAS Number: 9000-11-7)

Other denominations: sodium carboxy methyl cellulose, CMC

Origin

Modified celluloses come from cellulose, which is dispersed in an alkaline solution to form alkaline cellulose (Figure 1). This alkaline cellulose is forced to react with different chemical compounds that form different methyl ethers (table 1).

Composition and chemical structure

Modified celluloses come from the cellulose molecule which is a linear homopolisaccharide composed of glucose units linked by β (1-4) bonds, repeated "n" times (this "n" will determine the degree of polymerization (DP) of the molecule). Each glucose has three hydroxyl groups that can be replaced giving certain characteristics to the solution, this will determine the degree of substitution (DS) molecule.

The properties of the modified cellulose will be determined by:

- Degree of polymerization of cellulose chain (DP): a higher degree of polymerization will give a higher viscosity of the medium.

- Degree of substitution of the cellulose chain (DS): each glucose has 3 free hydroxyl groups that can be replaced, which will influence the solubility of the molecule and its subsequent functionality. According to the substitution realized, different types of ethers are obtained (table 2).

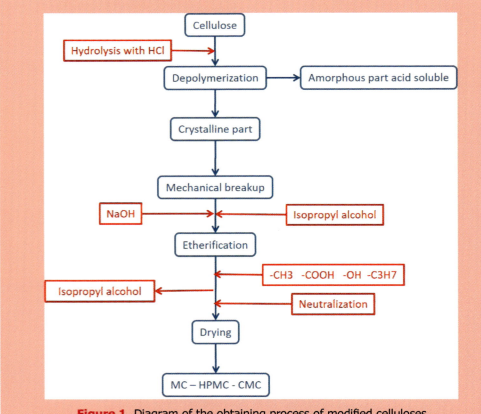

Figure 1. Diagram of the obtaining process of modified celluloses

Modified celluloses

E-461
E-464
E-466

Table 1. Modified celluloses obtaining process

Methyl ether	Reactive	Substituent group	Molecule obtained
E-461 Methyl cellulose	Methyl chloride	Methyl group (-CH$_3$)	
E-464 Hydroxypropyl methyl cellulose	Methyl chloride + propylene oxide	Methyl group (-CH$_3$) + propyl group (-C$_3$H$_7$)	
E-466 Sodium carboxy methyl cellulose	Sodium monochloroa-cetate	Sodium acetate group (-CH$_2$COONa)	

Table 2. Non-ionic and ionic modified celluloses

NON-IONIC	IONIC
E-461 Methyl cellulose (MC)	E-466 Carboxy methyl cellulose (CMC)
E-464 Hydroxypropyl methyl cellulose (HPMC)	

> ℹ Unlike other hydrocolloids, and as a particular feature, non-ionic ethers of cellulose (MC and HPMC) form a gel by heating them at a certain temperature and maintain this gelation as long as they maintain the temperature.

Properties of solutions

Solubility and solutions preparation

The properties of methyl cellulose will be determined by the degree of polymerization (DP) and the degree of substitution (DS). Table 3 shows the solubility characteristics of methylcellulose according to its degree of substitution, whose maximum value is 3.

Methyl cellulose is soluble in cold water, but it is difficult to disperse it. This requires very strong shear force and, therefore, there is a high incorporation of air (Figure 2). It is recommended to premix the methyl cellulose with other ingredients to improve its dispersion in cold water. Once dissolved, it increases the viscosity of the medium. Methyl cellulose is very stable to the attack of microorganisms and enzymes.

To avoid the formation of foam, the

Handbook of hydrocolloids

MC can be dispersed in hot conditions. Due that it is insoluble in hot conditions, to ensure its functionality the solution must be cooled to hydrate and become soluble (Table 4).

The same happens for the HPMC molecule.

Factors affecting the properties of solutions

- **Temperature:** by heating a methyl cellulose solution it is observed that viscosity decreases up to a certain temperature (incipient gel temperature (IGT)) from which a reversible gel is formed. When it finishes heating, gel melts and the solution gets its initial viscosity (Table 5).

- **pH:** this hydrocolloid is stable in a wide range of pH (3-11).

- **Ionic charge:** the presence of ions makes increase the viscosity of the medium (Figure 3).

Figure 2. Methyl cellulose with air and defoamed

Table 3. Solubility of the MC as a function of the degree of substitution

Degree of substitution	Solubility
< 1.6	Soluble in alkali
1, 6-2, 0	Maximum solubility in water
> 2.0	Soluble in organic solvents

Table 4. Solubility and solutions preparation / +++ (high) ++ (middle) + (low)

Solubility in cold water	Solubilization temperature	Premix	Shear force	Air incorporation	Increase viscosity of medium	Precipitates with alcohol
+++	Cold	++	+++	+++	Depends on DP	No

Modified celluloses

Table 5. Factors affecting the properties of solutions

Factors	Effects on viscosity	Observations
Molecular weight / polymerization degree	Directly proportional	
Concentration	Directly proportional	
Temperature	Inversely proportional to IGT	More temperature, less viscosity. From IGT it gels
pH	Stable over a wide range (3-11)	
Ionic charge	Directly proportional	
Mechanical work	Pseudoplastic behaviour	More shear force, less viscosity.

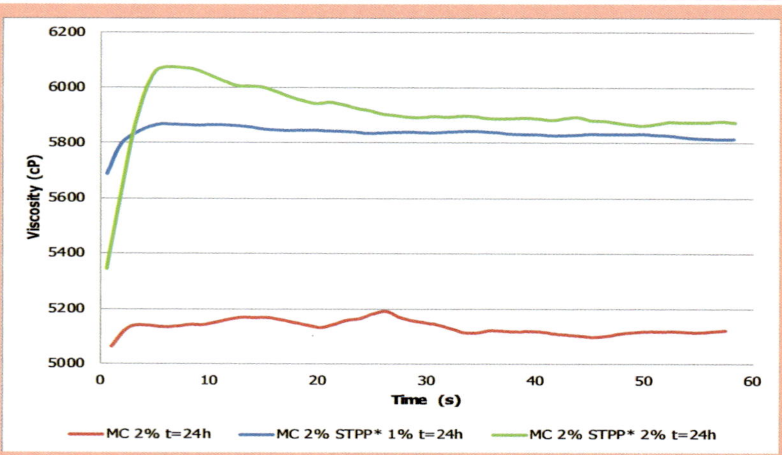

Figure 3. MC viscosity graph with different concentrations of ion (rotational viscometer (50 rpm, 60 s, spindle R6) /* Tripolyphosphate pentasodium

Functionality

Gelation mechanism

Methyl cellulose forms a thermoreversible gel from the incipient gel temperature (IGT) which will vary depending on the degree of substitution of the molecule. By stopping the heating, the gel will melt leaving a solution that will be recovering the initial viscosity. Gelation temperature and melting temperature are different, with a hysteresis loop (Figure 4). At higher concentration, the starting gelation temperature increases, and at the same time, the melting temperature will also vary (Table 6).

It is noticed that at high concentrations of methyl cellulose (≥4%) the obtained gel after heating does not melt, even though it loses some of strength, giving a thermostable gel (figure 9 and 10).

Handbook of hydrocolloids

Table 6. Gelation and melting temperatures of the MC as function of the concentration

Concentration (%)	Gelation temperature (ºC)	Melting temperature (ºC)
0.5	50	44
1	55	40
2	59	32
4	59	It maintains the gelling

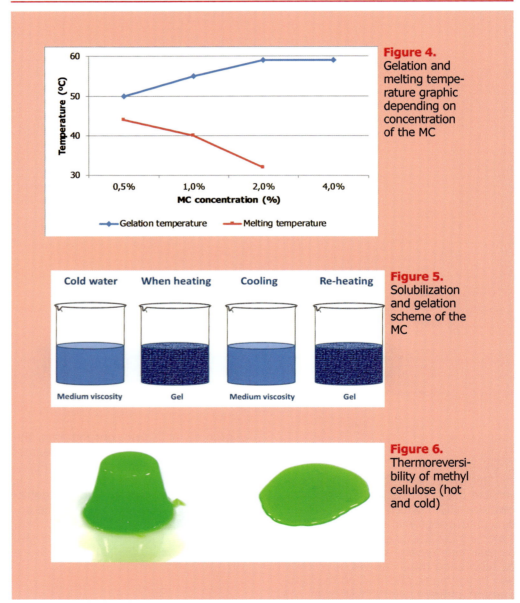

Figure 4. Gelation and melting temperature graphic depending on concentration of the MC

Figure 5. Solubilization and gelation scheme of the MC

Figure 6. Thermoreversibility of methyl cellulose (hot and cold)

Modified celluloses

Form of preparation

Methyl cellulose is dispersed in cold water and heated. In first place there is a decrease in the viscosity but when it arrives to incipient gel temperature (which varies depending on the concentration and DS) will begin to form a thermoreversible gel (Figures 5 and 6).

Factors affecting the properties of gels

- **Temperature:** it must reach temperatures higher than 50ºC.

- **Ionic charge:** it decreases the starting gelation temperature. Therefore, at the same concentration of MC, gelation occurs faster (Figure 7 and Table 7).

- **Alcohol:** the MC solubilization with alcohol does not make foam and they do not form a gel when heating, i.e., the viscosity stays both cold and hot conditions.

- **Propylene glycol:** by mixing it with the MC it does not form foam and they do not gel when heating, they lose viscosity, and precipitates.

Characteristics of gels

Depending on the MC concentration, different types of gels can be obtained (Figure 8 and Table 8).

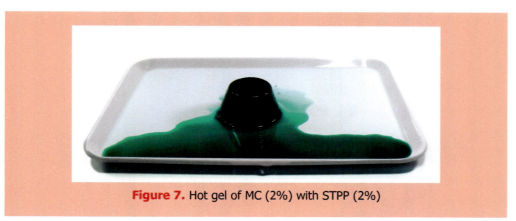

Figure 7. Hot gel of MC (2%) with STPP (2%)

Table 7. MC gelation temperature depending on the ionic charge

Concentration (%)	Gelation temperature (ºC)
2% MC	59
2% MC + 2% STPP	50

Table 8. Characteristics of gels

	High concentrations of MC	Low concentrations of MC
Gel type	Weak	Rigid
Syneresis	No	Yes
Temperature stability	Thermostable at T>59ºC and 4% MC	Thermoreversible at T> 50ºC and 0,5% MC

Handbook of hydrocolloids

Figure 8. Thermoreversible gel of methyl cellulose and detail of cutting

Figure 9. Methyl cellulose thermostable gel and detail of the cut in hot conditions

Figure 10. Thermostable methyl cellulose gel and detail of cutting once cooled

Synergies or incompatibilities
Not described.

Examples of applications in food industry

- Gelling agent in hot conditions for baked or fried products.

VEGGIE BURGER

Ingredients: soya beans, water, soy protein, onion, tomato concentrate, sunflower oil, salt, sugar, wheat fiber, onion juice concentrate, gelling agent (**methyl cellulose**)

Modified celluloses

E-461
E-464
E-466

- Stabilizer in foams and emulsions of toppings and ice cream.

- Film formation in frying processes, reduces the penetration of oil and improves the adhesion of coats.

LEMON SORBET

Ingredients: Water, sugar, lemon juice concentrate (15%), glucose syrup, flavourings, stabilizers (**methyl cellulose**, pectins, locust bean gum), acidity regulator (citric acid)

- Thickener in cold conditions for sauces, pasta.

- It helps give structure to gluten-free bakery and pastry products.

- Foaming agent.

ONION DELICES

Ingredients: Onion (40,5%), breadcrumbs (wheat flour, water, wheat gluten, salt, vegetable fat, yeast, raising agents (sodium bicarbonate, disodium diphosphate), emulsifier (mono - and diglycerides of fatty acids)), water, sunflower oil, wheat and corn flour, modified starch, sugar, thickeners (**methyl cellulose**, guar gum), gelling agent (sodium alginate), salt

---

CREAMY CARROT FOAM

Ingredients: carrot (95%), thickener (**methyl cellulose**), salt

---

> ℹ Modified celluloses have a particular commercial denomination, despite not being official, widely used in food industry. This nomenclature classifies them in: A, E, F and K. Under the letter A there is included the MC and under the E, F and K, the HPMC.
>
> Then, it is indicated its viscosity measured in a 2% solution. This is described with a number indicating the value and a letter indicating units (C: hundreds, M: miles). For example:
>
> - A4C indicates that the product is a MC with a viscosity at 2% of 400 cp.
>
> - F50M is an HPMC with a viscosity at 2% of 50000 cp.

> In the *Regulation (EU) No 1129/2011 from the Commission of 11 November 2011 amending Annex II to Regulation (EC) No 1333/2008 of the European Parliament and of the Council to establish a list of food additives of the Union* appears the E-463 or hydroxypropyl cellulose with CAS number: 9004-65-3.
> The HPC is obtained from cellulose which, previously dispersed in an alkaline solution, forms the alkaline cellulose.
> The behaviour of HPC is very similar to the HPMC. This alkaline cellulose is made to react with propylene oxide giving HPC.
> In this manual we will not explain HPC since its use is very uncommon in food.

E-464 HYDROXYPROPYL METHYL CELLULOSE

Properties of solutions

Solubility and solutions preparation

As is the case of the MC, the properties of the HPMC will be determined by the degree of polymerization (DP) and the degree of substitution of the cellulose (DS).

The HPMC has very similar characteristics to the MC. It is soluble in cold water, but it is difficult to disperse, also requires enough mechanical work that produces high air incorporation (Figure 11). The premixing of the HPMC with other ingredients improves its dispersion in cold water. Once dissolved, it increases the viscosity of the medium (Table 9).

Factors affecting the properties of solutions

- **Temperature:** by heating a hydroxypropyl methyl cellulose solution it is observed that viscosity decreases up to a certain temperature (incipient gel temperature or IGT) from which a reversible gel is formed. When it finishes heating, gel melts and the solution gets its initial viscosity (Table 10).

- **pH:** this hydrocolloid is stable in a wide range of pH (3-11).

- **Ionic charge:** the presence of ions produces a decrease in the viscosity of the medium (Figure 12).

Table 9. Solubility and solutions preparation / +++ (high) ++ (middle) + (low)

Solubility in cold water	Solubilization temperature	Premix	Shear force	Air incorporation	Increase viscosity of medium	Precipitates with alcohol
+++	Cold	++	++	++	Depends on DP	No

Modified celluloses

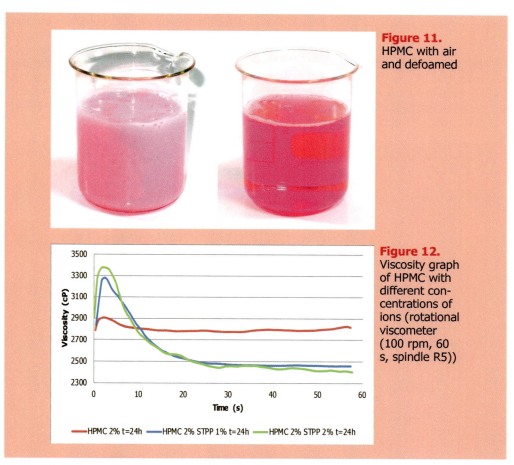

Figure 11. HPMC with air and defoamed

Figure 12. Viscosity graph of HPMC with different concentrations of ions (rotational viscometer (100 rpm, 60 s, spindle R5))

Table 10. Factors affecting the properties of solutions

Factors	Effects on viscosity	Observations
Molecular weight / Polymerization degree	Directly proportional	
Concentration	Directly proportional	
Temperature	Inversely proportional to IGT	More temperature, less viscosity. From IGT it gels or increases the viscosity
pH	Stable over a wide range	
Ionic charge	Inversely proportional	
Mechanical work	Pseudoplastic behaviour	More shear force, less viscosity.

Functionality

Gelation mechanism

The HPMC forms thermoreversible gels from the incipient gel temperature (IGT), which will vary depending on the degree of substitution of the molecule. When stopping to apply heat, and at a given temperature, the gel will melt and the solution will recover its initial viscosity. The HPMC also has a hysteresis loop by having different gelation and melting temperatures. A higher concentration of HPMC, greater starting gelation temperature, and also varying the melting temperature (Figure 13 and Table 11).

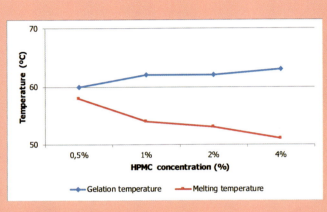

Figure 13. Gelation and melting graphic temperature depending on concentration of the HPMC

Figure 14. Solubilization and gelation scheme of the HPMC

Figure 15. Thermoreversibility of the HPMC (hot and cold)

Modified celluloses

E-461
E-464
E-466

Table 11. Gelation and melting temperatures of the HPMC as function of the concentration

Concentration (%)	Gelation temperature (°C)	Melting temperature (°C)
0.5	60	58
1	62	54
2	62	53
4	63	51

Form of preparation

The HPMC is dispersed in cold water. If heats the dissolution a decrease of viscosity is first observed but from a (IGT) temperature (which varies depending on the concentration and DS) will begin to form a thermoreversible gel (Figures 14 and 15).

Factors affecting the properties of gels
- **Temperature:** it must reach temperatures higher than 60ºC.

- **Ionic charge:** it decreases the starting gelation temperature. Therefore, at the same concentration of HPMC, gelation occurs faster (Table 12).

- **Alcohol:** when mixed with the HPMC they do not make foam and the solution has a higher viscosity than the same concentration of HPMC dispersed in water. They don't make a gel.

- **Propylene glycol:** by mixing it with the HPMC they don't make foam, they precipitate. When heated the precipitate gels giving a transparent gel.

Characteristics of gels

The HPMC forms thermoreversible gels (Table 13).

Synergies or incompatibilities

Not described.

Examples of applications in food industry
- Thickener in cold conditions for sauces, pasta.

- Stabilizer in foams and emulsions of toppings and ice cream.

- Film formation in frying processes, reduces the penetration of oil and improves the adhesion of coats.

Table 12. The HPMC gelation temperature depending on the ionic charge

Concentration (%)	Temperature Gelation (ºC)
2% HPMC	60
2% HPMC + 2% STPP	30

Table 13. Characteristics of gels

	HPMC
Gel type	Soft
Syneresis	Yes
Temperature stability	Thermoreversible at T> 60ºC and 0,5% HPMC

- Gelling agent in hot baked or fried products

> **HAM CROQUETTES**
>
> Ingredients: reconstituted skimmed milk, water, wheat flour, margarine, bread crumbs, onion, cured ham (4,3%), vegetable protein, potato starch, cooked ham, salt, flavourings, gelling agents (carrageenan, locust bean gum, guar gum), egg albumin, spices, thickener (**hydroxypropyl methyl cellulose**), preservative (potassium sorbate)

- It helps give structure to gluten-free bakery and pastry products.

> **GLUTEN-FREE TOASTED BREAD**
>
> Ingredients: potato starch, corn starch, rice flour, margarine (palm fat, coconut fat, water, sunflower oil, salt, emulsifiers (mono and di-glycerides of fatty acids), natural flavouring), whole rice flour, sugar, rice syrup, corn flour, chicory fiber, thickener (**hydroxypropyl methyl cellulose**), yeast, salt, natural flavouring

- Foaming agent.

E-466 CARBOXY METHYL CELLULOSE / CELLULOSE GUM

Properties of solutions

Solubility and solutions preparation

Carboxy methyl cellulose is soluble in both cold and hot conditions and it do not precipitate in the presence of alcohol. Hydration is very fast; for that reason, right away that comes in contact with the aqueous medium particles get wet forming large lumps, and they need the use of intense mechanical agitation to its correct dissolution. As a result, there is an incorporation of air in the mix, although not as important as in the case of other cellulose modified as the MC or the HPMC. The degree of substitution is directly related to the hydration, the CMC with higher DS has an increased dissolution rate (Table 14).

Factors affecting the properties of solutions

While the viscosity of a solution of CMC is basically related to the degree of polymerization (DP), there are some factors that can affect it (Table 15):

- **Ionic charge of the medium:** as it is an anionic polyelectrolyte the presence of cations interferes by decreasing the viscosity of the solution and, in the case of divalent cations such as calcium, even precipitating. In a medium with large amount of calcium ions, for example milk, the use of a sequestrant is necessary to prevent the precipitation of the CMC. Also, as in the case of microcrystalline cellulose colloidal grade, requires a homogenization stage, known as "activate the CMC" in the dairy industry.

- **Temperature:** by heating the solution, the viscosity of the medium will decrease, but it will return to the starting point after cooling (Fi-

Modified celluloses

gure 16). It is necessary to check that the temperature is not excessive or a very prolonged warming, since it can cause a loss of viscosity by hydrolysis of the chains of CMC.

Functionality

The CMC increases the viscosity of the medium (Figures 17 and 18) but do not form a gel, in contrast to non-ionic ethers of cellulose.

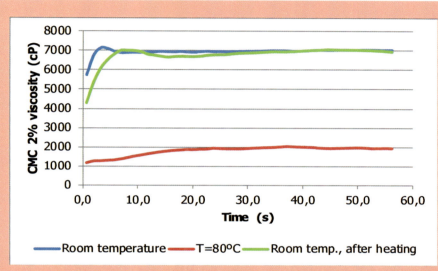

Figure 16. Viscosity of a solution of CMC 2% in cold conditions, during heating at 80°C and after cooling it again to room temperature (rotational viscometer, 50 rpm, 60 s, spindle R5)

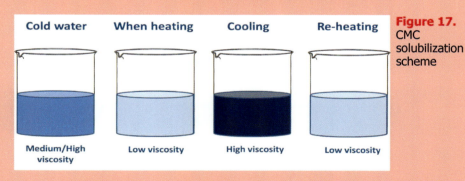

Figure 17. CMC solubilization scheme

Figure 18. CMC before and after heating at 80°C

Table 14. Solubility and solutions preparation / +++ (high) ++ (middle) + (low)

Solubility in cold water	Solubilization temperature	Premix	Shear force	Air incorporation	Increase viscosity of medium	Precipitates with alcohol	Ion sequestrant
+++	Cold	++	+++	+	It depends on the DP	No	++

Table 15. Factors affecting the properties of solutions

Factors	Effects on viscosity	Observations
Molecular weight / polymerization degree	Directly proportional	
Concentration	Directly proportional	
Temperature	Inversely proportional	If the heating is excessive, the decrease of viscosity may be permanent
pH	It is stable over a wide pH range	
Ionic charge	Is it affected by cations decreasing viscosity	With divalent cations, it precipitate, it needs a sequestrant in dairy medium
Mechanical work	Pseudoplastic behaviour	It decreases the viscosity as it increases the shear force

Modified celluloses

> In the literature we find described a case of gelation of the CMC which occurs in the presence of trivalent cations in the medium, in particular, aluminum salts. This application currently has no interest for application in the food industry, since the use of aluminum salts is not allowed.

- CMC + native starch: decreases the syneresis and increases the hardness of the gels formed with native starches, as shown in Figure 19 where a 8% cornstarch gel has increased significantly the hardness by adding 0,5% of CMC after the solubilization of the starch.

- CMC + MCC: interaction made by CMC and microcrystalline cellulose produces the commercially known as colloidal grade MCC, whose functionality and features are described in the chapter 14: E-460 Microcrystalline cellulose.

- CMC interaction with proteins: in diary or vegetable drinks CMC forms a stable complex that protects the denaturation and precipitation, during heat treatment, of proteins near their isoelectric point.

Synergies or incompatibilities

- CMC + locust bean gum / CMC + guar gum: the use of CMC in combination with one of these two hydrocolloids produce an increase in the viscosity of the mix higher than expected from the sum of the two hydrocolloids viscosities separately.

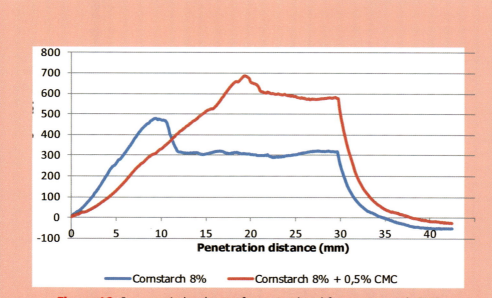

Figure 19. Increase in hardness of cornstarch gel for synergy with CMC

Handbook of hydrocolloids

Examples of applications in food industry

- Emulsion stabilizer.

COCONUT MILK

Ingredients: coconut extract (56%), water, stabilizer (**carboxy methyl cellulose**), emulsifier (mono and diglycerides of fatty acids), acidity regulator (citric acid).

- Humectant to prevent dryness in bakery products.

BRIOCHE

Ingredients: wheat flour (55%), water, sugar, canola oil, eggs, yeast, salt, wheat gluten, emulsifier (mono and diglycerides of fatty acids), stabilizer (**carboxy methyl cellulose**), flavouring, colour (beta carotene), acidity regulator (ascorbic acid), milk proteins.

- Stabilizer in ice cream: prevents the recrystallization of lactose.

CHOCOLATE ICE CREAM

Ingredients: whole milk, cream, glucose syrup, alkalized cocoa, skimmed milk, buttermilk, stabilizers (microcrystalline cellulose, **carboxy methyl cellulose**, carrageenan), emulsifier (mono and diglycerides of fatty acids).

- Film forming to facilitate peeling of hot dog's sausages

VIENA CHEESE SAUSAGES

Ingredients: mechanically separated meat from chicken and turkey, meat and bacon from pork, water, cheese (10%), salt, pork plasma, stabilizers (diphosphates, **carboxy methyl cellulose**, carrageenan), spices, flavourings, antioxidant (sodium erythorbate) and preservative (sodium nitrite).

- Stabilizer in low pH milk drinks.

Table 16. Summary table of celluloses

Cellulose characteristics	MCC	MC	HPMC	CMC
Viscosity in cold	Only that of colloidal grade by effect of CMC	Yes	Yes	Yes
Gel type	It does not gel	Very strong and with syneresis	Softer than MC or just increase of viscosity	No gel, increase of viscosity
IGT temperature	It does not apply	> 50ºC	> 60ºC	It does not apply
Melting temperature	It does not apply	< 45ºC	< 60ºC	It does not apply

16 E-1204 PULLULAN

E-1204 Pullulan (CAS Number: 9057-02-7)
Other denominations: pullulane

PULLULAN

Handbook of hydrocolloids

Origin

Pullulan is an exopolysaccharide produced by the fungus *Aureobasidium pullulans*, known as "black yeast" as it also produces melanin (figure 1).

Although *Aureobasidium* microorganism are commonly used for the production of pullulan, there are other genera and species that also produce it (table 1).

Composition and chemical structure

Pullulan is a linear hydrocolloid formed by the repetition of maltotriose units formed by 2 glucoses linked α(1-4) and another linked α(1-6), giving the molecule a shape we would recognize as stairs (figure 2). There sometimes may be ramifications. The final group is usually a maltotetraose. This linear structure of glucoses resembles as those of maltodextrins or some types of starches.

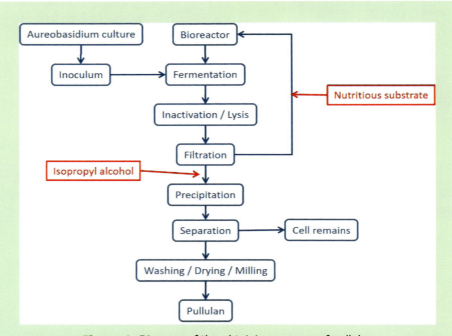

Figure 1. Diagram of the obtaining process of pullulan

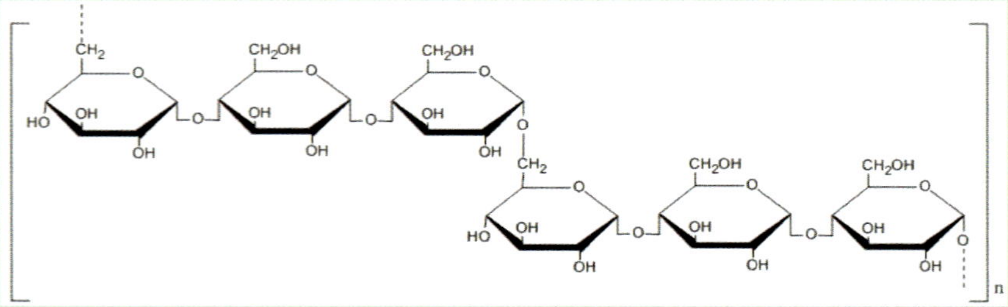

Figure 2. Pullulan molecule

Table 1. Pullulan-producing microorganisms

Microorganism	Reference
Aureobasidium pullulans	Bauer (1938); Cooke (1959); Leathers (2003)
Tremella mesenterica	Fraser and Jennings (1971)
Cytaria harioti	Olive (1986); Waksman (1977)
Cytaria darwinii	Olive (1986); Waksman (1977)
Cryphonecteria parasitic	Corsaro (1998); Delben (2006); Forabosco (2006)
Teloschistes flavicans	Reis (2002)
Rhodototula bacarum	Chi and Zhao (2003)

Properties of solutions

Solubility and solutions preparation (Table 2)

Pullulan is soluble in cold water, therefore, there's no need to heat the solution (figure 3).

Factors affecting the properties of solutions (Table 3)

- **pH:** pullulan solutions are very stable even in extreme pH values and do not suffer hydrolysis unless, in addition to an extremely low pH, the solution is brought to boiling.

- **Temperature:** pullulan solutions are stable to long and extended heat procedures, including processes of sterilization.

Table 2. Solubility and solutions preparation / +++ (high) ++ (middle) + (low)

Solubility in cold water	Solubilization temperature	Premix	Shear force	Air incorporation	Increase viscosity of medium	Precipitates with alcohol
+++	cold	+++	+++	+	+	Yes

Table 3. Factors affecting the properties of solutions

Factors	Effects on viscosity	Observations
Molecular weight / polymerization degree	Directly proportional	
Concentration	Directly proportional	
Temperature	Inversely proportional	An increase in temperature causes a decrease in viscosity
pH	Very stable in a wide range of pH (2-11)	Warming extended at pH <3 can reduce the viscosity by partial hydrolysis of the molecule
Ionic charge	Maintains viscosity at high salt concentration	
Mechanical work	Pseudoplastic behaviour	It decreases the viscosity as it increases the shear force

Functionality

Pullulan provides low viscosity solutions. However, it is very adhesive and easily forms films that are quite impermeable to oxygen, even at low concentrations.

Synergies or incompatibilities

They have not been described.

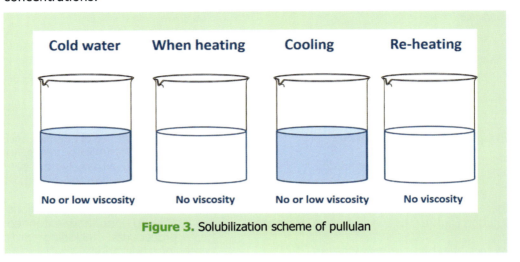

Figure 3. Solubilization scheme of pullulan

Examples of applications in food industry

- Formation of films that are quite impervious to oxygen in nut coating.

SALTY PEANUTS

Ingredients: Peanuts, sunflower oil, salt, stabilizer (**pullulan**)

- Formation of adhesive films for seeds on the surface of breads or pastries.
- Formation of edible films for protection of fruit.
- In pharmacy, for their ease of forming edible films against bad breath.

FRESH STRIPS FOR BAD BREATH

Pullulan, menthol, sucralose, potassium acesulfame, oleate, glycerin, polysorbate 80, carrageenan, thymol, eucalyptol, methylsalicylate, locust bean gum, propyleneglycol, xanthan gum and flavouring.

Identification and GRAS status

Hydrocolloid	GRAS status (year of report)	CAS number	E number
Alginate	1973	9005-38-3	E-401
Agar	1973	9002-18-0	E-406
Carrageenan	1973	11114-20-8 (kappa) 9062-07-1 (iota) 9064-07-1 (lambda)	E-407
Locust bean gum	1972	9000-40-2	E-410
Guar gum	1973	9000-30-2	E-412
Tragacanth gum	1972	9000-65-1	E-413
Gum arabic	1973	9000-01-5	E-414
Xanthan gum		11138-66-2	E-415
Karaya gum	1973	9000-36-6	E-416
Tara gum		39300-88-4	E-417
Gellan		71010-52-1	E-418
Konjac		37220-17-0	E-425
Cassia gum		51434-18-5	E-427
Pectin	1977	9000-69-5	E-440
Microcrystalline cellulose	1973	9004-34-6	E-460
Methyl cellulose	1973	9004-67-5	E-461
Hydroxypropyl methyl cellulose	1973	9004-65-3	E-464
Carboxy methyl cellulose	1973	9000-11-7	E-466
Pullulan		9057-02-7	E-1204

BIBLIOGRAPHY

1. ALVARADO, J. D. y AGUILERA, J. M. (2001). *Métodos para medir propiedades físicas en industrias de alimentos*. Zaragoza: Acribia.

2. ANTON, R.; BARLOW, S. et al. (2004). *Opinion of the Scientific Panel on food additives, flavourings, processing aids and materials in contact with food (AFC) on a request from the Commission related to Pullulan PI-20 for use as a new food additive*. EFSA Journal [en línea]. Disponible en: https://www.efsa.europa.eu/en/efsajournal/pub/85 [noviembre 2015].

3. Asociación Española de Normalización y Certificación. *Análisis sensorial. Guía general para la selección, entrenamiento y control de catadores y catadores expertos*. UNE EN ISO 8586:2012. AENOR, 2012.

4. Asociación Española de Normalización y Certificación. *Análisis sensorial. Vocabulario*. UNE 87001:1994. AENOR, 1994.

5. BARNES, H. A.; HUTTON, J. F. y WALTERS K. (1993). *An introduction to rheology*. Amsterdam: Elsevier Science Publishers, B. V.

6. BOURNE, M. C. (2002). *Food texture and viscosity: Concept and measurement* (2a ed.). San Diego: Academic Press.

7. BRISHTAR LABORATORIOS, C.A. *Goma tragacanto* [en línea]. Disponible en: http://www.bristhar.com.ve/tragacanto.html [agosto 2015].

8. CARGILL. *Pectin functionality* [en línea]. Disponible en: https://www.cargillfoods.com/ap/en/products/hydrocolloids/pectins/functionality/index.jjs [enero 2016].

9. CHAPLIN, M. *Hydrocolloids and gums* en Water Structure and Science [en línea]. Disponible en: http://www1.lsbu.ac.uk/water/hydrocolloids_gums.html [octubre 2015].

10. CUBERO, N; MONFERRER, A y VILLALTA, J. (2002). *Aditivos alimentarios*. Barcelona: Mundi Prensa.

11. FOSTER, T.J. (2010). *Hydrocolloids Structure and Properties. The building blocks for structure* [en línea]. Disponible en: http://www.stepitn.eu/wp-content/uploads/2009/04/pm18_Unilever_Foster_1.pdf [agosto 2015].

12. FURIA, T.E. (1981) *Handbook of food additives* (2a ed.). Florida: CRC Press.

13. GULREZ, S.; AL-ASAF, S. (2011). Hydrogels: Methods of Preparation, Characterisation and Applications. En A. CARPI, *Progress in Molecular and Environmental Bioengineering* (cap. 5) [en línea]. Croacia: INTECH. Disponible en: http://www.intechopen.com/books/progress-in-molecular-and-environmental-bioengineering-from-analysis-and-modeling-to-technology-applications/hydrogels-methods-of-preparation-characterisation-and-applications [diciembre 2015].

14. HERNÁNDEZ, M.J. (2006). *Reología y textura*. Apuntes del curso telemático sobre hidrocoloides. Organizado por BDN Ingeniería de Alimentación S.L. Barcelona.

15. IKA WORKS. *Application information of food and beverages* [en línea]. Disponible en: http://www.ika.com.my/appFood.html [octubre 2015].

16. KRAMER, A. (1964). *Definition of texture and its measurement in vegetable products. Food Technology*. 18, 304.

17. LARDMON, E. (1977). *Laboratory Methods for Sensory Evaluation of Foods. Publication Canada Department of Agriculture*. Nr. 1637. Ottawa: Canada Department of Agriculture.

18. MULLER, H.G. (1969). *Mechanical properties, rheology and haptaesthesis of food. Journal of texture studies*. 1 (1), 38-42.

19. NUSSINOVITCH, A. (1997) *Hydrocolloid applications: Gum Technology in the Food and Other Industries*. London: Blackie Academic & Professional.

20. PHILIPS, G.O y WILLIAMS, P.A. (2000). *Handbook of hydrocolloids*. Florida: CRC Press.

21. SÁNCHEZ, C; RENARD, D; ROBERT, P; SCHMITT, C y LEFEVRE, J. (2002). *Structure and Rheological properties of acacia gum dispersions. Food Hydrocolloids*. 16, 257-267.

22. SARATHCHANDIRAN, I y SURESH KUMAR, P. (2014). *Characterization and standardization of gum karaya. International Journal of Biopharmaceutics.* 5 (2), 142-151.

23. SMITH, J. (1991). *Food additives user's handbook*. London: Blackie Academic & Professional.

24. STEFFE, J. F. (1992). *Rheological methods in food process engineering*. Michigan: Freeman Press.

25. WHISTLER, R.L.; BEMILLER, J.M. (1993). *Industrial Gums* (3a ed.). San Diego: Academic Press Inc.